Japan's Whaling

JAPANESE SOCIETY SERIES
General Editor: Yoshio Sugimoto

Gender and Japanese Management
Kimiko Kimoto

Philosophy of Agricultural Science: A Japanese Perspective
Osamu Soda

A Social History of Science and Technology in
Contempory Japan, Volume 3
Shigeru Nakayama and Kunio Goto

Japan's Underclass: Day Laborers and the Homeless
Hideo Aoki

A Social History of Science and Technology
in Contemporary Japan, Volume 4
Shigeru Nakayama and Hitoshi Yoshioka

Scams and Sweeteners: A Sociology of Fraud
Masahiro Ogino

Toyota's Assembly Line: A View from the Factory Floor
Ryoji Ihara

Village Life in Modern Japan: An Environmental Perspective
Akira Furukawa

Social Welfare in Japan: Principles and Applications
Kojun Furukawa

Escape from Work: Freelancing Youth and the Challenge to Corporate Japan
Reiko Kosugi

Gender Gymnastics: Performing and Consuming Japan's Takarazuka Revue
Leonie R. Stickland

Poverty and Social Welfare in Japan
Masami Iwata and Akihiko Nishizawa

Japan's Whaling: The Politics of Culture in Historical Perspective
Hiroyuki Watanabe

Health and Social Disparity: Japan and Beyond
Norito Kawakami, Yasuki Kobayashi and Hideki Hashimoto

Social Stratification and Inequality Series

Inequality amid Affluence: Social Stratification in Japan
Junsuke Hara and Kazuo Seiyama

Intentional Social Change: A Rational Choice Theory
Yoshimichi Sato

Constructing Civil Society in Japan:
Voices of Environmental Movements
Koichi Hasegawa

Deciphering Stratification and Inequality: Japan and beyond
Yoshimichi Sato

Social Justice in Japan: Concepts, Theories and Paradigms
Ken-ichi Ohbuchi

Gender and Career in Japan
Atsuko Suzuki

Status and Stratification: Cultural Forms in East and Southeast Asia
Mutsuhiko Shima

Globalization, Minorities and Civil Society:
Perspectives from Asian and Western Cities
Koichi Hasegawa and Naoki Yoshihara

MODERNITY AND IDENTITY IN ASIA SERIES

Japan's Whaling

The Politics of Culture in Historical Perspective

Hiroyuki Watanabe

Translated by

Hugh Clarke

Trans Pacific Press

Melbourne

First published in Japanese as *Hogei mondai no rekishi shakaigaku* in 2006 by Tōshindō.
Trans Pacific Press, PO Box 164, Balwyn North, Victoria 3104, Australia
Telephone: +61 (0)3 9859 1112 Fax: +61 (0)3 9859 4110
Email: tpp.mail@gmail.com
Web: http://www.transpacificpress.com

Copyright © Trans Pacific Press 2009

Designed and set by digital environs, Melbourne, Australia. www.digitalenvirons.com

Printed by BPA Print Group, Burwood, Victoria, Australia

Distributors

Australia and New Zealand
UNIREPS
University of New South Wales
Sydney, NSW 2052
Australia
Telephone: +61(0)2-9664-0999
Fax: +61(0)2-9664-5420
Email: info.press@unsw.edu.au
Web: http://www.unireps.com.au

USA and Canada
International Specialized Book
Services (ISBS)
920 NE 58th Avenue, Suite 300
Portland, Oregon 97213-3786
USA
Telephone: (800) 944-6190
Fax: (503) 280-8832
Email: orders@isbs.com
Web: http://www.isbs.com

Asia and the Pacific
Kinokuniya Company Ltd.

Head office:
3-7-10 Shimomeguro
Meguro-ku
Tokyo 153-8504
Japan
Telephone: +81(0)3-6910-0531
Fax: +81(0)3-6420-1362
Email: bkimp@kinokuniya.co.jp
Web: www.kinokuniya.co.jp

Asia-Pacific office:
Kinokuniya Book Stores of Singapore Pte., Ltd.
391B Orchard Road #13-06/07/08
Ngee Ann City Tower B
Singapore 238874
Telephone: +65 6276 5558
Fax: +65 6276 5570
Email: SSO@kinokuniya.co.jp

ISSN   1443–9670 (Japanese Society Series)

ISBN   978–1–876843–75–5 (Hardcover)
ISBN   978–1–876843–69–4 (Paperback)

Cover illustration: The tail fluke of a humpback whale in waters off Bondi Beach, as it migrates south to the Southern Ocean, as Japanese whaling boats are also heading to the Southern Ocean to begin their whale hunting season, 27 September 2007. Courtesy of *The Sydney Morning Herald*. Picture by Dallas Kilponen.

Contents

Figures

Tables

Translator's Preface

The whaling issue generates a great deal of negative press for the Japanese government. To Japan's closest friends – its political allies and trading partners in North America, Europe and Oceania – what is seen as the Japanese government's stubborn insistence on perpetuating a dying, unprofitable industry is nothing short of exasperating. It is difficult for them to comprehend why Japan should be prepared to tarnish its international image for the sake of a relatively small but vociferous whaling lobby. Despite concerted efforts in Japan to create a stable market for whale meat, many Japanese, particularly among the younger generation, have never tasted it and have no apparent desire to do so. Most of those who *have* tasted it would never dream of making whale meat their preferred substitute for beef, pork or chicken. Why, then, is the Japanese government so vigorous in advocating a return to commercial whaling? Is it because they see western opposition to whaling as an unwarranted attack on traditional Japanese culture? Or is it, perhaps, through a fear that the ban on whaling might be the thin edge of the wedge of a prohibition on the exploitation of other wildlife resources like, say, tuna?

As Watanabe Hiroyuki points out in this book, the position the Japanese government takes in current debates in the International Whaling Commission can only be understood in the historical context of Japanese whaling from the end of the nineteenth century. The author also brings a degree of balance to the debate by demonstrating that the government's current position simply reflects one of several different relationships between human beings and whales that were prevalent in Japan in the early modern period. Watanabe shows how the current focus on whaling for meat production stems from the Japanese government's sponsorship of whaling as a kind of 'national industry' and its exploitation of whales in its colonial waters around the Korean peninsula with Russian whaling vessels confiscated during the Russo-Japanese war of 1904–05.

Watanabe also refutes the claim, so often voiced in the IWC, that whaling is traditional Japanese culture. He contests the definition of culture proffered by certain Japanese and western scholars and questions the legitimacy of branding the localised practices of a small number of coastal communities as 'Japanese tradition.'

Importantly, the author also reminds his Japanese readership, and informs readers of this English edition, about the early opposition to large-scale commercial whaling in many parts of Japan. In some areas, Watanabe tells us, the taking of whales was prohibited, as it was believed that these huge marine mammals were incarnations of Ebisu, the god of fishermen, who blessed coastal communities with bountiful catches of sardines.

The resistance to whaling culminated in the riot and burning of Tōyō Whaling's station at Same village in Aomori prefecture in 1911. Watanabe's painstaking research on the evidence from the trial of the fishermen involved in the Same riot has provided a graphic account of the impact the introduction of whaling had on the lives and emotions of a community of sardine fishermen. He also considers the implications of life experience and 'common sense' in the context of social movements theory, relating the data from Same to the work of the likes of Alain Touraine and Alberto Melucci.

In his attack on the 'whaling culture' argument Watanabe points out that modern Japanese whaling is based on Norwegian technology introduced at the end of the nineteenth century. In the early period, the harpoon gunners, and often the captains, of whaling catcher boats were Norwegians and the crews were comprised of Korean and Chinese as well as Japanese seamen. Japanese whaling was thus very much a multi-national enterprise operating in an international whaling arena. Any continuities with the traditional net whaling, dating back to the seventeenth century, of certain parts of Kyushu, Shikoku and Wakayama were largely either superficial or coincidental.

The crux of Watanabe's argument is that through the historical contingencies of Japan's colonial rule over the Korean peninsula, the Russo-Japanese war and imperialistic expansionism, a plurality of relationships between human beings and whales in Japan has been reduced to a single relationship – catching whales for their meat. To demonstrate his point, Watanabe also analyses whaling discourse that provides a rationale for overfishing and the Japanese attitude to the conservation of wildlife that emphasises the need to protect species utilised as an economic resource.

As a Japanophile, but one who shares his countrymen's general distaste for whaling (admittedly a little hypocritical, given

Australia's involvement in whaling throughout the nineteenth century), I am delighted that through this translation I have been able to help disseminate a Japanese scholar's new angle on the whaling debate and demonstrate that not all Japanese are fanatical supporters of commercial whaling. Just as Watanabe has shown us that it is inappropriate for the Japanese government to generalise the ideas and practices of a small number of people living in a few scattered villages into 'Japanese whaling culture,' so too have we in the West been guilty of tarring all Japanese with the same anti-whaling brush.

Recently, the Japanese chapter of Greenpeace hit the headlines with the accusation that some members of the crew of the scientific whaling vessel Nisshin Maru had been smuggling whale meat ashore for their own private (or commercial?) use (*The Japan Times*, 16 May 2008, p.1). Further, as whales gradually begin returning to Japanese coastal waters, whale-watching is becoming a lucrative new attraction for the tourist industry.

In translating this book into English I encountered a number of problems I had not envisaged. Given the inconsistency in the English names for many fish and animal species, I decided that the English translation should have the scientific name appended to the English gloss. It was only then that I discovered there is often disagreement among specialists regarding the scientific nomenclature; there are even differing opinions about the number of existing whale species. The scientific names given in the text are therefore open to question. For the most part, they have been supplied by Watanabe Hiroyuki himself. Some I have taken from encyclopaedias or the Internet.

Another problem concerns the Japanese translation of books and articles originally written in English or French. Where Watanabe used only the Japanese version, reference to the Japanese translation is included in the footnotes. This is necessary as a record of the author's source, since the Japanese translation often abbreviates sections of the original or, conversely, includes additional material for the benefit of the Japanese readership. For readers of the English version, however, where I have been able to identify the original English (or a published English translation of a French work), I have used that for quotations and the pagination of in-text citations.

A further difficulty arises from the slightly different usage of quotation marks in Japanese and English. Whereas in English, quotation marks around a word or phrase indicate that the usage is unusual – either that the words carry a connotation of irony, sarcasm or scepticism; or, that the discussion concerns the word itself rather than the concept it conveys. In Japanese, on the other hand, in addition to these uses, key concepts or words carrying emphasis are often placed in quotation marks. For example, Watanabe invariably places quotation marks around '*bunka*,' the Japanese word for 'culture;' the inference being that he himself is not happy with the concept of culture in sociological argument. While I have tried to follow Watanabe's usage as long as it does not create a jarring effect in English, sometimes I have felt it necessary for the sake of English expression to omit the quotation marks of the Japanese original. I would like to take this opportunity to apologise to Watanabe for taking this liberty, though I do not think the meaning or the tone of the text has been compromised by the decisions I have made in this regard.

Although this book contains a wealth of material that should appeal to the general reader with an interest in whaling or in Japanese culture, it is fundamentally a scholarly work – based as it is on the author's Ph.D. thesis – with a very strong theoretical component and numerous citations to an impressive corpus of academic literature. I am sure specialists in Japanese whaling or in historical sociology will also welcome Watanabe's comprehensive bibliography and detailed index.

Finally, I would like to thank Watanabe for the patience and meticulous attention he devoted to the correction of several earlier drafts of my translation and for teaching me so much about Japanese whaling. My thanks also go to the editor, Professor Yoshio Sugimoto of Trans Pacific Press, for guiding me through this challenging yet stimulating project. Every effort has been made to eliminate errors and infelicitous expression from the English translation; any shortcomings that remain are entirely my responsibility.

Hugh Clarke
Sydney, July 2008

Preface

I wonder what Japanese readers today feel when they hear the word '*kujira*' (whale). No doubt the first thing to spring to mind would be the so-called whaling issue. We can certainly say that this is question of deep concern to Japan. This is obvious from the regularity of the news coverage of the whaling issue at the time of the annual meeting of the International Whaling Commission. But what we read in the press, particularly since the introduction of the moratorium on whaling in 1987, is about the structure of opposing views countries hold over whaling and how, within that structure, Japan takes a strong stance in advocating its desire to resume whaling. Recently, in addition, the argument that 'whales consume large quantities of marine resources that are beneficial to human beings' has been more forcefully pursued and has been given wide coverage.

Or older readers might recall how whale meat used to appear on the dinner table or be served up in school lunches. It may be that these memories and experiences have led them to be vaguely persuaded by the argument that 'the ban on whaling is cultural bullying by Europe and America.'

Ultimately, the aim of this book is to put forward one opinion regarding this whaling issue. It does not, however, seek to evaluate the facts from the viewpoint of the natural sciences. The approach I have chosen is to elucidate the history of the relationship between whales and human beings. Why the emphasis on history; you may ask. The reason is precisely to avoid the kind of structural confrontation mentioned above. To achieve this, we now need to step back and calmly assess the situation, rather than getting excited about forcefully setting out the validity of our own position. By calm assessment I mean that we should try to adopt a relative stance and objective perspective; tasks accompanied by a degree of introspection.

In this task of introspection I have endeavoured to look back over the events of the past, giving each very careful consideration. In addition, in the course of my examination I have made a point of applying an analytical framework from the discipline of sociology.

In elucidating the history of whaling there may be times when my conclusions differ from the experiences and recollections of particular individuals' associations with whales. This is not to say that I deliberately set out to oppose those individuals' evidence, but we must realise that there are often unseen circumstances behind those experiences and recollections. What I do reject is not the individual experiences and recollections, but rather the conglomerated experience and memory that have come to be constructed by the various forces that have appropriated the stories of each of those individuals. And when it comes to the issue of whaling, this conglomerated experience and memory is called Japanese 'tradition' and Japanese 'culture.'

Consequently, it is true to say that this book is neither pro- nor anti-, whaling. It is a call for us to work together to consider the future connections between whales and people, by breaking away from this amalgamated position and placing the facts, one by one, together with their interpretations, in front of every individual concerned with the issue.

Conventions adopted in this book

1. References are given in parentheses (…) as in-text citations in the order of author, followed by the year of publication and then the page number. A list of all works cited, arranged alphabetically on the author's surname, is included at the end of the book. Where the author is unclear, the title of the work is given instead. Web pages are listed under the title of the article followed by the Internet URL. These are treated separately and are listed at the end of the book grouped together and arranged in order according to the chapter in which they are cited. Reference to the Japanese translations of books in other languages is included in the Notes and in the Bibliography, but not in the in-text citations unless the original source was not available for confirmation.

2. I have omitted the phonetic glosses to Sino-Japanese characters, *furigana*. Where I feel explanation about a quotation or a cited work is warranted I include my author's comments in angle brackets […].

3. The English translation follows the traditional order, family name followed by given name, for Japanese, Korean and Chinese names. For the romanisation of Korean names the official revised romanisation is used unless we are aware of an individual's preference for an alternative Anglicised spelling. For common Korean place names that have become firmly established in English spelling, e.g. Pusan, the traditional spelling is used in favour of the Republic of Korea's official romanisation, Busan. Japanese names are transcribed in the modified Hepburn romanisation and Chinese names are in the official Chinese pinyin romanisation.

4. A glossary of Japanese whaling company names is included as an appendix to the English translation. The full Japanese name is given the first time a company is mentioned, but for subsequent citations the short reference form is used. See the Appendix for literal translations of the Japanese company names.

Prologue: The Aim of this Book

Introduction

Among the works of Ōsaka Keikichi, the detective fiction writer who emerged on the literary stage in 1932 at the young age of twenty and died of illness during the fighting in Luzon in 1945, there is one with the title *The Pod Stayed Put* (*Ugokanu Geigun*) (Ōsaka, 1936 [2001]).

The story goes like this. It is one year after the Norwegian-style whaling ship the *Hokkai Maru* sank in a storm. A woman who lost her partner, the harpoon-gunner on the ill-fated ship, has begun working in a bar in Nemuro to support her infant child. But through the glass in the door of the bar she sees the form of the harpoon-gunner she had assumed to be dead. She leaves the room and runs after her partner. The gunner explains that he has been longing to see his child, but that someone is pursuing him. He pleads with the woman to bring the baby and run away with him. Later, when she returns to his hiding place with the baby, she finds the gunner lying cruelly impaled on a whaling harpoon. With his final dying breath he says, 'it was the captain of the *Kushiro Maru*.'

The *Kushiro Maru* was the other of the two catcher boats licensed to the company that owned the *Hokkai Maru*. The police immediately set out to search for the *Kushiro Maru* only to discover that it had already left port. When they went on to question the boss of the company it seemed as if he had something to hide.

Keen to get to the truth of the matter, the director of the Fisheries Research Station, along with several police officers, boarded a surveillance vessel of the Fisheries Bureau and left the harbour in pursuit of the *Kushiro Maru*. When the surveillance ship sent out a radio signal that it had sighted a pod of whales, the *Kushiro Maru* appeared and began whaling. The surveillance ship seized the opportunity to pull in close to the *Kushiro Maru* and issue an order for the vessel to heave to. When the police and the director of the research station boarded the ship they found there the captain and the entire crew of the *Hokkai Maru*, all of whom should have been dead.

The director of the Fisheries Research Station deduced the truth of the incident as follows. The company that owned the *Hokkai Maru* had claimed that the ship had sunk so it could build another vessel to replace it and so surreptitiously run three ships instead of the two for which it held whaling licences, thereby increasing the company's productivity. Of course, the *Hokkai Maru*, which was supposed to have sunk and was now sailing under the false name of *Kushiro Maru*, could not put into port anywhere near Nemuro. But the harpoon gunner eventually grew homesick and jumped ship to be with his wife and child. So, the captain on the fake *Kushiro Maru* murdered him to prevent the secret from leaking out.

Of course this is a work of fiction. But the story has a lot to teach us. Namely, that Japanese whaling ships formerly employed Norwegian harpoon-gunners and that, to protect whale stocks, the Japanese government restricted the total number of whale catchers to thirty vessels. Further, at the time of the story, whaling was carried out within the territorial waters of 'Japan' as it was then, that is to say, mainly along the east coast of Hokkaidō and around the Korean Peninsula, then a Japanese colony. And in a major underlying theme of the work we learn what significance the harpooning of a *whale calf* (*kokujira*) had for the whaling industry (Ōsaka, 1936 [2001]).

When news of the sinking of *Hokkai Maru* gets around the port, there are whispers among the older folk about the legend of the curse of the whales visited upon those who kill a whale calf. In addition, the primary reason those on the surveillance vessel order the fake *Kushiro Maru* to heave to is that they have just seen it harpoon a young whale, an act expressly forbidden by law. Further, we are told that when a calf is harpooned the other whales in the pod remain in the area – hence the title of the story – and are easier to catch.

This is why, we are told, the captain of the fake *Kushiro Maru* had ordered the runaway harpoon-gunner to shoot whale calves. The story ends with the author's explanation that parent whales travel more slowly when they are with their calves and that they would never abandon their offspring as the harpoon-gunner, who had gone missing for a whole year, had done.

But, as we shall see later when we look closely at the history of whaling, at the time the number of licensed ships was restricted to thirty vessels there was no ban on the taking of whale calves or of hunting whales accompanied by their young. On the contrary, whales with calves were considered the most suitable target for whaling vessels. Nevertheless, we cannot help but be impressed when reading this work, by the author's powers of imagination in depicting the fluctuating emotions of those who confront the gigantic vitality of whales, only to have to kill them. Was the harpoon gunner's unbearable desire to see his child perhaps exacerbated by the fact that he was required to shoot whale calves in the line of his daily duties? Isn't it likely that the people around the port would also come to see the gunner's murder as another case of the curse of the whales?

Be that as it may, I get the feeling that these kinds of images are no longer admissible in what we read and hear about 'the whaling issue.' This is not to say that human emotions are being ignored. Rather, the emotions have been organised, fixed and systematised to become aggressively intertwined into the whaling debate. We have lost the ability to imagine how others feel, and consequently are no longer able to expand our own rigid, single-minded emotions.

In a sense, it is to achieve this expansion of the emotions that we must now begin our analysis of the whaling issue.

The whaling issue today

What then is the current state of the whaling issue? The moratorium on whaling, agreed at the 1982 meeting of the International Whaling Commission (henceforth abbreviated to IWC) and brought into force at the end of 1987, is, at the time of writing (2005), coming to the end of its eighteenth year of operation (refer to Table 5.1 for an overall summary of the whaling issue). Throughout this period the international confrontation surrounding whaling has been maintained, with for example, the declaration of a whale sanctuary in the Southern Ocean[1] on the one hand, and the withdrawal from the IWC or the resumption of whaling by the so-called whaling nations, Iceland and Norway, on the other.

Nevertheless, the whaling industry has already reached the point where it would be very difficult for it to return to the large-scale production it formerly enjoyed. Indiscriminate killing of whales, particularly since the advent of Norwegian-style whaling – the system which employs a harpoon with an explosive head propelled from a harpoon gun mounted on the bow of a steam vessel – in the latter half of the nineteenth century, up until the 1960s, has meant that some of the species once hunted commercially have all but been driven to extinction. Take, for example, the largest creature on this planet, the blue whale (21 to 27 metres in length and weighing from 100 to 120 tons) (Carwardine 1995).[2] At the height of the whaling industry (1930–31 season) in the major whaling grounds of the Antarctic Ocean thirty thousand of these animals were taken in a single year. It is estimated that there are now no more than five hundred of them swimming around the southern oceans (Katō 1991), (Kawabata 1995), (Ōsumi 1994). For this reason, the majority of these are now protected and the main focus of the debates in the IWC has been the minke whales, said to number around one million (Sakuramoto, Katō *et al.* 1991), which, being the smallest of the baleen whales, had largely escaped the excessive hunting of the past.[3]

It is common knowledge that Japan, as a whaling nation, is right in the firing line of the debates in the IWC. Since the introduction of the moratorium on whaling, Japan has been carrying out scientific whaling to collect a wide range of biological data. Surveys have been carried out in the Southern Ocean since 1987 and the northwest Pacific Ocean since 1994. Further, according to a current report of July 2003, scientific whaling of minke whales in the Southern Ocean and minke, Bryde's, sei and sperm whales in the northwest Pacific is about to begin ("The whaling situation at home and abroad, July 2003," *http://www.jfa.maff.go.jp/whale/document/brief_ explanation_of_whaling_jp.htm*). And the Japanese government has continued to lobby for a minke whale quota, based on its assertion that Japan's small-type coastal whaling[4] can be considered to fall under the category of aboriginal subsistence whaling,[5] i.e. whaling by indigenous populations for food or ritual purposes.[6]

In this way, whaling has become an issue because our excessive inclination to exploit a specific wild-animal resource, has led not only to a reduction in the number of whale habitats, but has also damaged the ecological system. Consequently, the question of whaling should probably be treated as an environmental issue. The present volume represents my attempt to examine the whaling issue from the perspective of historical sociology. In so doing, and bearing in mind Japan's special position in regard to the whaling issue, I shall be focusing my attention on the debate within Japan and the controversy over Japan's present and past involvement in whaling.

Analytical perspective

Considering then the debate in Japan, we can identify a point of culmination in the arguments of Akimichi Tomoya and Morita

Katsuaki (Akimichi 1994: Morita, 1994). Actually, these two researchers reach roughly the same conclusion. Both emphasise the diversity of relationships[7] between whales and human beings and incorporate the findings of anthropological research. They seek a new relationship between whales and people incorporating 'the sense of the supernatural – the feelings of fear and awe towards wild animals and the tension involved in hunting and fishing,' (Akimichi 1994: 203) or 'the accumulated knowledge and experience of living with and utilising nature' (Morita 1994: 420) of, for Akimichi, the indigenous peoples and 'non-westerners' living on the Pacific rim and, for Morita, the people at present living in certain areas of Japan. Consequently, both of them not only dismiss the western anti-whaling argument that takes the form, in a sense, of personifying whales, but also reject the pro-whaling, nationalist argument, Japan has been promulgating since the advent of the moratorium, that the eating of whale meat is 'a tradition' of the 'Japanese people.' In particular, this unprecedented denial of the nationalist argument brought a valuable new perspective to the debate.

Akimichi and Morita's rejection of both the anti-whaling and pro-whaling arguments, as we have seen above, may be seen as a rebuttal of what they considered political motivation in the use of these claims in the whaling debate, both within and outside the IWC. Further, both scholars assert that the relationship between people and whales, particularly in the context of whaling, is to be treated as 'culture.' To crush political dogmatism, Morita demands that 'culture' be scientific in the sense that it can be empirically verified (Morita 1994: 415–6). On the other hand, Akimichi is negative in his appraisal of the use of science in the whaling debate, saying that science is an ideology that sees whales as a 'resource' and merely reduces them to numbers (Akimichi 1994: 201–2). Consequently, by describing 'culture' in the form of an ethnography, he 'seeks to write the story of whales and people that is neither scientific nor political' (Akimichi 1994: i).

However, one cannot imagine any piece of folkloric or anthropological research representing 'culture' as being completely free of political concerns. In fact, recent reflective folkloric and anthropological research (e.g. Clifford 1988 [2003]; Clifford and Marcus eds 1986 [1996]; Iwatake 1996; Marcus and Fischer 1986 [1989]) addresses the political aspect of academic representations of 'culture.' That is to say, researchers intentionally determine what constitutes a certain 'culture' in the context of specific historical and social conditions and in relation to other 'cultures.' Similarly, the identity of a people based on that 'culture' is not fixed, but is politically constructed. And it has become clear that the activity of researchers portraying 'culture' has, since the beginning of the early modern period, played a fixed role in the construction of nation states and peoples and continues to do so in the present day whenever these are reconstructed.

It seems to me that, in the light of these recent trends in folklore and anthropology, we need to carefully examine the idea of 'culture' proposed by Akimichi and Morita. Can we really say, for example, that treating the relationship between whales and human beings under the rubric of 'culture' is not in the least political?

In fact, Akimichi participated in the survey of small-type coastal whaling carried out under the direction of the anthropologists M. M. R. Freeman and Takahashi Jun'ichi. Morita, for his part, relies completely on both the published report of this research (Freeman 1988 [1989]) and the work of Takahashi himself (Takahashi 1987; 1991; 1992). This means, therefore, that it is necessary for us to extend our consideration to these works.

A more detailed discussion will follow, but Freeman *et al.* states that 'Anthropologists generally use the word "culture" to mean "shared knowledge" transmitted through a socialization process from one generation to the next' (Freeman 1988: 28). It takes this idea further with the statement,

> A whaling culture, such as that described here, may be defined as the shared knowledge of whaling transmitted across generations.
>
> This shared knowledge consists of a number of different socio-cultural inputs: a common heritage and world view, an understanding of ecological (including spiritual) and technological relations between human beings and whales, special distribution processes, and a food culture.
>
> The common heritage found in Japan's whaling culture is based on a long historical tradition. In this respect, it is primarily focussed on time, in that it relates myths, folk tales, legends and other narrative events concerning whales and whaling. (Freeman 1988: 75 [1989] 165–6).

In this way, Freeman sets out to expound on the existence of what he calls 'whaling culture.'

Further, Takahashi ends up employing a manipulative definition of culture as,

The integrated system of consolidated knowledge, skills and social organisation required for people to seek, find, obtain, prepare/process, then distribute and consume, the resources of the ecological environment in which they live (Takahashi 1992: 19).

And, he goes on to say,

When we can recognise a phenomenon in which a particular group of people adopts a unique lifestyle based on whaling activity that is organically linked to various social, economic, technical and spiritual aspects of the group, we can call it 'whaling culture' (Takahashi 1992: 21).

Having said that, Takahashi goes on to compare, mainly from the aspects of modes of hunting and processing of the carcass, types of whaling that were formerly or are currently still practised in Japan, namely the large-type coastal whaling employing large steam-powered vessels or factory-ship whaling based on a ship with facilities for processing the whale carcass, usually for the production of whale meat and whale oil, with the net-whaling method established at the end of the seventeenth century, in which an organisation known as the *kujiragumi* (whaling group) caught and dispatched whales by driving them into nets and spearing them with harpoons (Takahashi 1992: 90–116). Consequently, he claims that in Japanese 'whaling culture,' that is to say, 'the culture centred on the subsistence activity (or industry) of utilising the whale resource within the marine environment on the edge of the northwest Pacific we call Japanese coastal waters,' whether it be whaling using nets, large-type coastal, or factory-ship whaling, 'despite superficial structural differences, these exhibit a great number of commonalities (...) most of these common features are found in those fundamental areas that characterise Japanese whaling' (Takahashi 1992: 28–29, 113).

Further, in order to demonstrate this point, Takahashi first cites the fact that there are clear continuities in the division between catching and processing activities, seen in differences in techniques, knowledge and the membership of the local and kinship groups of those involved. As added support for his claim, he cites the reciprocity between the operators (whaling group or whaling company) and the regions where fishing grounds and processing plants are located, together with the continuity and conservatism of techniques and manufacturing processes based on a stable 'tradition' of whale meat consumption (Takahashi 1992: 90–116).

In these representations of 'culture' and 'whaling culture' articulated in Freeman *et al.* and by Takahashi it is clear that these researchers base their arguments on history. It follows then, to examine their discourse, we must clarify the historical facts concerning the relationship between whales and people in Japan. And clarifying this point means showing that there is political motivation behind any claim that, once an object of an event has been construed as 'culture,' there is an historical continuity linking past and present. We must be aware that this narrative is one of the type substantially posited when a particular group, in order to define itself as a nation state or a people, claims, that a particular phenomenon, is a fundamental and unchanging attribute of that group, despite the fact that the phenomenon might have undergone a variety of changes over time. (See, for example, Iwatake ed. 1996: 20–1, 32–40).

Of course, we cannot deny that the writing of history invariably includes politics. Certainly, indisputable 'facts' existed in the past. But the accumulation of these facts results in a history with a context (i.e. intent). Researchers, without actually introducing fallacious material, employ the methodology of 'deletion of the truth,' by excluding this or that fact from their representation.[8] Consequently, history can be seen as 'continual reconstruction of the past in the present' made by a researcher following a certain paradigm (Ueno 1997).

In short, since there is invariably a political implication in any representation, it boils down to the question of to what purpose or from whose point of view is this representation being made? So, we can say that my description of the history of our relationship with whales as it is set out in the following pages also expresses a particular political stance on the whaling issue.

This book, then, aims to elucidate the history of the relationship between whales and human beings, adopting the analytical perspective I have outlined above. In particular, I have decided to concentrate on the early modern, modern and contemporary periods of the relationship. My principal reason for doing so is because I believe, in considering the whaling issue today, we need to understand the nature of the relationship between whales and human beings in the context of a period in which human activity has emphasised economies of scale and rationalisation. Secondly, and I shall be dealing with this in more detail in subsequent chapters, previous research on the whaling industry has been hardly what one would call voluminous, so I think it is necessary for our deliberations for me to present the basic facts. In the concluding section, having clarified the historical context, I attempt a critique of the anthropological discourse described above in terms of 'whaling culture theory.' Finally, I give an overview of the policy debates on the whaling issue and append some thoughts on the future relationships between whales and human beings.

1 Workers and the Introduction of Technology

The issue

Locating the problem

The aim of this chapter is to explain the introduction of technology into the Japanese whaling industry from the end of the nineteenth century. In doing so, I pay particular attention to the micro level, dealing with the composition of the labour force and the whaling companies that were the conduits of the new technology.

As I mentioned in the prologue, the anthropologist Takahashi Jun'ichi discussed the process of the introduction of this technology, arguing from the viewpoint of his concept of 'whaling culture.' Takahashi, comparing net whaling with large-type coastal whaling and factory-ship whaling, mainly in the areas of catching and processing activities, identifies areas of similarity and argues from these that there is continuity in Japanese 'whaling culture' from net whaling to large-scale coastal whaling and factory-ship whaling. Further, we can say that Takahashi's belief [1] that 'despite the various historical changes, there exists an underlying fundamental and unchanging whaling culture,' has added support to the claim held widely within Japan in regard to the so-called whaling issue that 'whaling is traditional Japanese culture.' But can we really say that Takahashi's thinking is on target? Ultimately, this is the question I wish to consider in this chapter and I do so by examining the facts concerning the introduction of whaling technology that Takahashi uses in his comparison.

First, looking at the analytical perspective, I should like to demonstrate, when claims are made about 'culture,' particularly about 'traditional culture' from the viewpoint of folklore and anthropology, precisely which attributes within the frame are admissible in establishing the legitimacy of any such claim.

It seems that, in general, when we hear the word 'culture' we tend to think of it as falling within the sphere of 'artistic attainment,' especially with contemporary music or fine arts and not as a term we would normally apply to folklore or anthropology. According to James Clifford, however, since the nineteenth century the concepts of 'art,' and 'culture' in the anthropological sense, began to be fused, with each supporting the other, as artifacts created by human beings, particularly those man-made objects with which Europeans came into contact in the process of colonial expansion, were collected, classified and conserved.

In short, from the nineteenth century in Europe, the field of 'art,' which classified and collected so-called 'primitive art' and modernist painting which had been influenced by primitivism, and 'culture,' the collection and classification of all manner of human lifestyles from various parts of the world, came to be linked into a single system. In this process, anthropologists defined the various 'cultures' of the world as if each were an independent body like a growing organism and regarded all as having equal value. Consequently, 'culture' was seen as having wholeness (in which the various elements are organised into a coherent system) and continuity, together with an impulse for development (Clifford 1988 [2003]: 293–8, 433–6, *passim*).

On the strength of these observations, especially considering the definition of culture as having wholeness and continuity, I myself, when considering the legitimacy of any claim that something falls under what anthropologists or folklorists call 'traditional culture,' inevitably ask myself whether it exhibits a combination of the following two characteristics. The first is continuity with the past. A phenomenon that has arisen over the past few years cannot be called 'traditional culture.' Conversely, any claim that something is 'traditional culture' must be able to demonstrate its continuity with the past and come up with effective arguments to counter assertions (like those proffered in the anti-whaling discourse in the debate over the whaling issue) that it is a recent phenomenon. The other point is concerned with categorisation of 'us' and 'those who are not us.' For example, while we readily accept expressions like 'Japanese tradition' or 'Kyoto culture,' we would no doubt feel uneasy about the terms 'human traditions,' and 'world culture.' This tells us that what we claim to be 'traditional culture' is composed of an arrangement of elements that combine to form a single organic whole, but which in fact is formed on the basis of comparison with a collective 'other' separate from 'us.' Consequently, in regard to the phenomenon of Japanese whaling, whether or not we are justified in calling this 'culture' must depend on the categorisation of 'Japanese' as opposed to 'Western.'

Below I describe the process of introduction of technology, then go on to consider, in the light of these facts and the two characteristics I have described above, whether it is appropriate for us to speak of 'whaling culture.'

Chronological divisions

Before embarking on a description of the introduction of whaling technology, I should like to clarify the timeline of the development of the modern Japanese whaling industry and set down the composition and standing of the whaling workforce.

We can think of the process of development of the Japanese whaling industry in terms of the following time divisions. The first phase (up to 1896) saw the decline of net whaling and the introduction of American-style whaling. Net whaling under the *kujiragumi* (whaling group) system had been suffering from poor catches since the end of the eighteenth century. There is room for debate over what may have caused this decline, but, in addition to natural conditions like changes in climate and ocean currents, there is speculation that perhaps it may have been due to the fact that the American whaling fleet began to come into Japanese coastal waters from the 1820s (Morita 1994: 315–317). Further, the so-called 'great right whale shipwreck' (*ōseminagare*)[2] incident of Taiji in Wakayama prefecture hastened the demise of net whaling. In response to this situation, according to John Manjirō (Nakahama) (1860s), Fujikawa Sankei (1873) and Sekizawa Akikiyo (1887) there were various attempts to introduce American-style whaling, employing large sailing vessels as mother ships and chasing whales from rowing boats armed with harpoons or a whaling gun called a 'bomb lance' (See, for example, Morita 1994: 51–124, 259–313, 317–20 and Takahashi 1992: 61–70, 77–8). But despite these various attempts the method never became fully established (Ishida 1978: 35–6; 'The account of Sekizawa Akikiyo' 1897a: 35–6; 'The account of Sekizawa Akikiyo' 1897b: 38–9; Morita 1994: 324–6).

The second period, from 1897 to 1908, is the period of introduction of Norwegian-style whaling, i.e. the method in which whales are harvested using a harpoon with an explosive head, fired from a gun mounted on the bow of a steam-powered vessel. The establishment and commencement of operation of the En'yō Hogei Kabushiki Kaisha (hereafter abbreviated to En'yō Whaling)[3] and the Nagasaki Hogei Kabushiki Kaisha (Nagasaki Whaling) in Nagasaki in 1897 heralded the start of Norwegian-style whaling by Japanese whaling companies (Torisu 1999: 336–42).[4] (See Figure 1.1 below)–. With the founding, two years later in 1899, of the Nihon En'yō Gyogyō Kabushiki Kaisha (abbreviated below to Nihon En'yō Fisheries) and its successful launch of whaling operations in the coastal waters around the Korean Peninsula, the Norwegian method became the established form of Japanese whaling. Further, after the Russo-Japanese War several more Japanese whaling companies were set up and Norwegian-style whaling spread throughout Japan itself (Tōyō Hogei K.K. 1910: 192–268).

The third period from 1909–33 marks the era of the whaling monopoly under the Tōyō Hogei Kabushiki Kaisha (below abbreviated to Tōyō Whaling). Tōyō Whaling was born in 1909 through the amalgamation, centred around Tōyō Gyogyō Kabushiki Kaisha (hereafter shortened to Tōyō Fisheries), the successor to Nihon En'yō Fisheries, of the numerous whaling companies that had sprung up indiscriminately after the Russo-Japanese War. This company came to have twenty whaling stations, including those on the Korean Peninsula just prior to Japanese annexation and, in compliance with the thirty vessel limitation on the number of Norwegian-style whaling boats prescribed under a directive issued by the Ministry of Agriculture and Commerce (*Nōshōmushō*) in 1909 (see also Chapter 5), twenty whale catchers. Tōyō Whaling's monopoly continued from these beginnings until the end of the period (Tōyō Hogei K.K. 1910: 13–23, 268–80).

The fourth stage, 1934–41, saw the advent of factory-ship whaling and the growth of whaling subsidiary companies under large capital investment. In addition to the worldwide economic slump since the so-called Great Depression, the sudden fall in the whale oil price brought on by global overproduction, had impacted so hard on the management of Tōyō Whaling that it resolved in 1930, for the first time in the company's history, not to pay a dividend. Further, along with the decline in catches of the comparatively larger whales like the grey whale, the right whale and the blue whale, which were considered the most important 'resource,' there was concern over the rising costs that came with the need to move up to larger vessels as whaling grounds moved, year by year, further offshore. In these circumstances, in 1934 the Ministry of Agriculture and Forestry (*Nōrinshō*), issued a directive cutting a further five vessels from the previous limit, bringing the ceiling down to twenty-five. Then, in the same year, Tōyō Whaling was bought out by Nihon Sangyō Kabushiki Kaisha (Nihon Production Co.), and Nihon Hogei Kabushiki Kaisha (Nihon Whaling), later to become the whaling division of Nihon Suisan Kabushiki Kaisha (Nihon Fisheries), was founded (Baba 1942: 89–94; Kaiyō Gyogyō Kyōkai (Ocean Fishing Association) 1939: 93, 97–9). It was under Nihon Whaling in 1934 that the Japanese whaling industry first began factory-ship whaling in the Antarctic Ocean. Then, in 1936, as a countermeasure to this activity, Hayashikane Shōten founded Taiyō Hogei and embarked upon factory-ship whaling. At around the same time in 1937 it bought Tosa Hogei K.K. (Tosa Whaling), which had for some time been a subsidiary under Hayashikane Shōten's umbrella, and turned it into the company's whaling division. Further, Sumatora Takushoku K.K. (Sumatra Colonial Co.) soon followed suit and entered the whaling industry. After first bringing Ayukawa Hogei K.K. (Ayukawa Whaling) into its organisation, it established Kyokuyō Hogei K.K. (Kyokuyō Whaling), embarking on factory-ship whaling from 1938 (Baba 1942: 94–107).

Finally, the fifth period from 1942 to 1945 is delineated by the suspension of factory-ship whaling and the advent of whaling under control companies. Factory-ship whaling was curtailed in 1941 with the outbreak of the Pacific War and, with the enactment of the Fisheries Control Ordinance in 1943, Nihon Fisheries was subsumed into the Nihon Kaiyō Gyogyō Tōsei Kabushiki Kaisha (Nihon Kaiyō Fisheries Control Co.) and the Hayashikane group companies into Nishi Taiyō Gyogyō Tōsei Kabushiki Kaisha (Nishi Taiyō Fisheries Control Co.).

The whaling boats were conscripted as patrol boats and the factory ships as tankers or transport vessels. All six of the requisitioned factory ships, in particular, were destroyed and sunk during the war (Maeda and Teraoka 1952: 21–2, 33–5; Park 1995: 291; Morita 1994: 351).

In this chapter I shall be concentrating mainly on stages two and four of these five phases in the development of the modern Japanese whaling industry, that is to say on the processes of introduction of Norwegian-style whaling and factory-ship whaling *per se*. Moreover, in regard to the three-way division of whaling administration, up to the time of the implementation of the moratorium on commercial whaling in 1987, into factory-ship whaling, large-type coastal whaling and small-type coastal whaling, I shall not be dealing here with the development of small-type coastal whaling (Freeman *et al.*, 1988 [1989]: 23–5, Preface f.n. 2). Further, I shall be treating the large catcher boats used in factory-ship whaling and large-scale coastal whaling in parallel with the Norwegian-style whaling ships introduced at the end of the nineteenth century.

The composition and status of the labour force

Generally, ships' crews, with the captain at the head, are divided into deck, engine room, wireless, purser's office and medical/sanitary section staff (Tennensha Jiten Henshūbu 1963: 232). Of course, there are slight variations in the composition of crews on more specialised vessels like whaling ships. Of particular relevance here, as shown in Figure 1.2, is the fact that factory ships have processing staff and on catching vessels there are the harpoon gunners. Let us compare the composition of workers in these categories during the second and third periods of Japanese whaling. First, as we can see from looking at the situation on whale catchers (Figure 1.2 and Tables 1.1 and 1.2), the composition of crews is roughly the same. However, immediately after the introduction of Norwegian-style whaling, because of the unavailability of on-shore processing factories (Mishima 1899: 34–5),[5] flensers were required on board, but this need soon disappeared. Conversely, not long after, with the advent of wireless, additional crew were required as radio operators.

In addition, we have the following description of the organisation of a Tōyō Whaling station at the beginning of the third period.

> In our organisation of the flensing and processing operation, we have followed the course of a divided system of labour, assigning a small number of administrative staff and several dozen workers to perform all the necessary tasks, which are allocated according to the skills each operator possesses into, head flenser, flenser, butcher, oil extractor, processor, storeroom attendant, engineer, blacksmith and so on. These all cooperate harmoniously under the supervision of the administrative officers (Tōyō Hogei K.K. 1910: 114).

We can assume that the composition of the processing crew on a factory ship would have been similar to that described here.

On the basis of this, in the following description of the introduction of technology, in order to get to the underlying truth, I would like to focus consideration on the bearers of this technology and, following Takahashi's description, divide them into those involved in catching and processing activities respectively, that is to say, into seamen and process workers. Consequently, I shall not be treating the engine room or related sections, nor shall I be looking at the crews of ships involved in transport or other activities. Further, in the description below, in order to distinguish high-ranking crew with titles like 'officer' or leading workers designated 'chief' this or that, from those

Figure 1.1: Changes in Norwegian-style and factory-ship whaling in Japan (1897–1945)

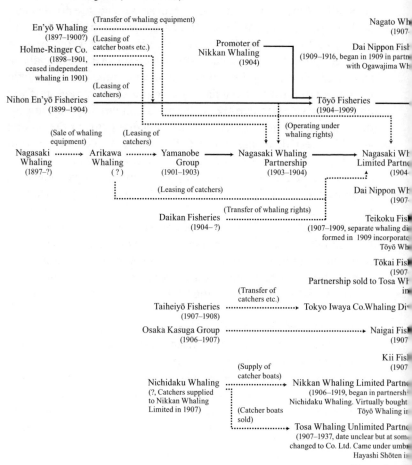

——— Purchase, amalgamation or change in company organisation

········· Lease, sale or supply of assests (whale catcher boats, whaling equipment etc.) and transfer of whaling rights.

Sources: Baba, Komao (1942), *Hogei*, Tennensha: 35–6, 89–122; Kyokuyō Hogei sanjūnenshi henshū iinkai ed. (1968), *Kyokuyō Hogei sanjūnenshi:* 214–32; Maeda, Keijirō and Teraoka Yoshirō (1952), Hogei, Nihon Hogei Kyōkai: 21–35, 43–9; Mishima, Tatsuo (1899), *Hogei shinron*, Sūzanbō; Nōrinshō Suisankyoku ed. (1939), *Hogeigyō*, Nōgyō to Suisansha: 3; Park, Koo-Byong (1995), *Jeungbopan Hanbando*

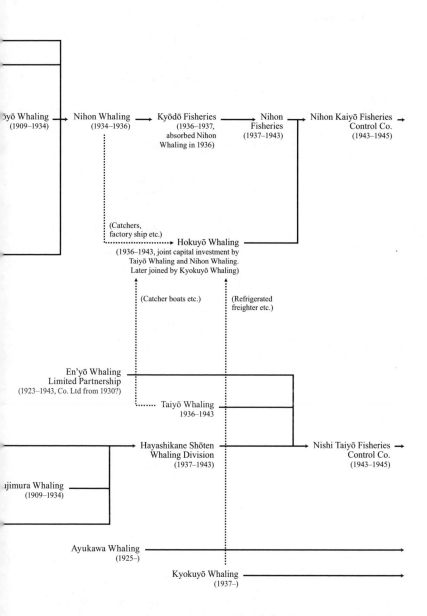

eonhae pogyeongsa, Pusan: Toseochulpan Mincheok Munhwa: 258–61, 282, 291; *Taiyō Gyogyō hachijūnenshi* hensan iinkai ed. (1960), *Taiyō Gyogyō hachijūnenshi*: 248–9, 285–300; Torisu, Kyōichi (1999), *Saikai hogei no shiteki kenkyū*, Kyūshū Daigaku Shuppankai: 336–42, 353–4; Tōyō Hogei K.K. ed. (1910), *Honpō no Noeruēshiki hogeishi*: 183–200.

Figure 1.2: Composition of crews in factory-ship whaling

Whale catcher

19 Crew members per vessel, total 171 (9 vessels). Made up of:

> 1 Harpoon gunner, 1 Captain, 1 Chief engineer,
> 1 Steersman, 1 Engineer, 1 Wireless operator,
> 1 Boatswain, 5 Seamen, 1 Chief stoker, 4 Stokers,
> 2 Stewards.

Whaling factory ship

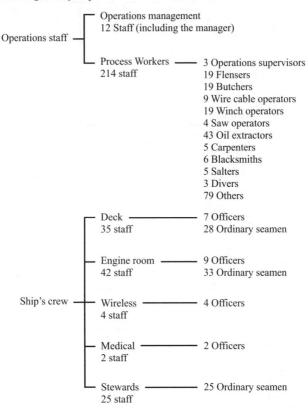

Operations staff
— Operations management
  12 Staff (including the manager)
— Process Workers — 3 Operations supervisors
  214 staff   19 Flensers
                   19 Butchers
                   9 Wire cable operators
                   19 Winch operators
                   4 Saw operators
                   43 Oil extractors
                   5 Carpenters
                   6 Blacksmiths
                   5 Salters
                   3 Divers
                   79 Others

Ship's crew
— Deck ————————— 7 Officers
  35 staff   28 Ordinary seamen
— Engine room ——— 9 Officers
  42 staff   33 Ordinary seamen
— Wireless ——————— 4 Officers
  4 staff
— Medical ———————— 2 Officers
  2 staff
— Stewards ——————— 25 Ordinary seamen
  25 staff

Total staff 334

Notes:

1. These figures are said to include the captain of the factory ship, but it is not clear under which category he has been listed.

2. The source gives the total number of process workers as 208. I presume this is a mistake and have corrected the figure to 214.

Source: Baba, Komao (1942), *Hogei*, Tokyo: Tennensha, 225–30.

Table 1.1: Crew of a catcher boat (2nd period)

	Number of people	Monthly salary (yen)
Captain	1	40
Harpoon gunner	1	100
Engineer	1	40
Purser	1	30
Flenser	5	12 (60/5)
Seamen	5	15 (75/5)
Stoker	4	15 (60/4)
Steward	1	25
Handy man (sic.)	1	10
Total	20	

Note: The total salary for flensers, seamen and stokers is given in the source material, along with a note to the effect that an average has been taken to arrive at the salary per person. From this we can assume that there was variation in payment for these jobs and that there were leaders within each of these occupational groups.

Source: Mishima, Tatsuo (1899), *Hogei shinron*, Sūzanbō, 32–3.

Table 1.2: Crew of a catcher boat (3rd period)

	Number of people
Captain	1
Harpoon gunner	1
Chief engineer	1
First engineer	1
Seamen	6
Stoker	2
Oiler	2
Cook	2
Boy	2
Total	18

Note: The source gives the total of 17. I presume this is a mistake and have corrected the figure to 18.

Source: Kawai, Kakuya (1924), *Zōho kaitei gyorōron*, Suisansha, 321–2.

workers positioned lower down in the hierarchy, I refer to the latter as 'general seamen' or 'general process workers.'

First, let us consider the organisation of net whaling. There is a considerable body of research on this subject, for example (Fukumoto 1978: 101–5; Hidemura 1952b: 67–75; Izukawa 1943: 105–13, 518–525, 1973c 125–33, 550–57); Morita 1994: 147–51; Kumano Taijiura hogeishi hensan iinkai 1969: 383–91). On the

Figure 1.3: Organisation of whaling groups (kujiragumi)

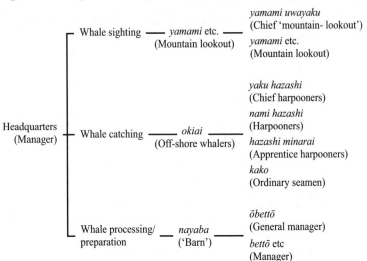

Note: This table is a very general illustration of the organisation of whaling groups. The names of the various positions vary considerably over time and from one whaling group to another.

basis of this research we can generalise the composition of labour as set out as in Figure 1.3. From this we can see that the whaling group (*kujiragumi*) was divided into three sections, around the headquarters (*honbu*), that was in charge of overall management. These were, the 'mountain lookout' (*yamami*), who directed catching activity by spotting whales from a high vantage point and directing boats to the catch, the 'off-shore' (*okiai*) who performed the actual catching of the whales, and the on-shore facility, known as the 'barn' (*nayaba*), for processing and the provision of equipment. Figure 1.4 shows the organisation of the Tosa (Kōchi prefecture) Ukitsu whaling group in 1795 and Figure 1.5 the same Tosa Tsuro whaling group in 1890 just before the introduction of Norwegian-style whaling.[6]

But, it is conjectured that the number of workers was probably actually higher than the figure indicated, as it is not clear in this data how many day labourers from surrounding districts were involved. In *Geishi* (Whale chronicle), written in the eighteenth century, 1,100 names are recorded for the combined membership of the two whaling groups of Taiji and Koza (both in present-day Wakayama prefecture) (Morita 1994: 151). And at the beginning of the nineteenth century in the Masutomi whaling group of

Figure 1.4: Organisation of the Ukitsu whaling groups (1795)

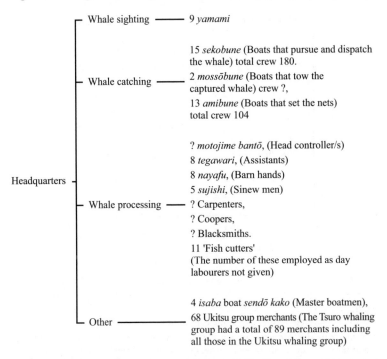

Source: Achikku myūzeamu (Attic Museum) ed. (1939). *Tosa Muroto Ukitsu gumi hogei shiryō* reprinted in *Nihon Jōmin Bunka Kenkyūjo* ed. [1973] *Nihon jōmin seikatsu shiryō sōsho dai jūnikan*, San'ichi Shobō, 524–5.

Ikitsukishima in Nagasaki it is recorded that at the Misaki fishing grounds off Ikitsukishima alone, there was a total of some nine hundred people including five hundred and eighty-seven regular members and the rest made up of day labourers (*Isantori ekotoba* 1832 [1829]: 285–92; Fukumoto 1978: [1993] 99–100; Morita 1994: 150–1). Further, we have a record from the mid nineteenth century that in the whaling group of Ogawajima (Saga prefecture) to capture a whale required eight hundred men and once the whale had been caught a further three hundred or more day labourers joined in (Hōshūtei 1840 [1995]: 367). From these accounts of net whaling it is clear that vastly greater numbers of workers were required to harvest and process each catch than with Norwegian-style whaling ships.

It is often pointed out that in general the status of workers in net whaling organisations reflected the social structure of the time. Actually, it is thought that the whaling group overlapped with

Figure 1.5: Organisation of the Tsuro whaling groups (1890s)

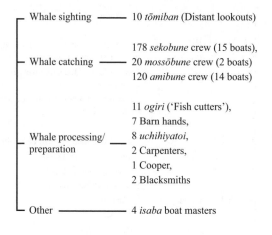

Total 364 people

Notes:

1. The source gives a total of 464 people, but this must be an error. Corrected to 364.

2. In regard to the duties of the *uchihiyatoi* (day labourers) the source material states. 'They carry out all miscellaneous tasks within the barn and assist the barn hands.Their work is little different from that of the barn hands.'

3. Apparently the *isaba* boats were provided by the whaling company to transport whale meat for sale in other areas, in cases where the brokers' bids were too low or the prospects for sale were poor.

Source: Tsuro Hogei K. K. ed. (1902), *Tsuro hogeishi:* Folio 63c (reverse side) – Folio 65c (reverse), 113c (front) – 113c (reverse), 142c (front).

the local community structure and that the headquarters (*honbu*) distributed daily rations of rice to the workers by way of payment (Hidemura 1952b: 98–104; Izukawa 1943: 85–6, 122–9 [1973c] 105–6, 142–9; Kumano Taijiura hogeishi hensan iinkai 1969: 383–7, 407, 409–13). And the positions in the headquarters, the *hazashi* (whose main job was to drive the harpoon into the whale) and most of the specialist processing positions in the barn (*nayaba*) were hereditary ones (Morita 1994: 152–3). In addition, it has been pointed out that those day labourers who were employed in the final butchering of the carcass, the flensing of the skin and preparation of the various detailed cuts of meat, were quite possibly members of the outcaste group known as *eta* (Tagami 1992; Yamashita 2004a: 183–92; Morita 1994: 152).

But this organisational structure changed in a variety of ways over time and from one whaling group to another with the penetration of the cash economy and increases in the scale of whaling group management. In Taiji, for example, labourers for whale catching activity came to be employed from outside the immediate vicinity, from as far away as Sugariura (in the area of present day Owase city in Mie prefecture) (Kumano Taijiura hogeishi hensan iinkai 1969: 450–2). And in the case of northwest Kyūshū too, where the whaling industry as represented by the Masutomi whaling group was of a comparatively large scale, in the same way, they looked beyond the area of the local community, in this case employing people from Bingo Tajima (on the Inland Sea in Hiroshima prefecture) for the manufacture of nets and ropes and to act as crew on net-setting boats (*sōkaisen*) (Hidemura 1952b: 70–2, 75–82; Oyamada 1832 [1829]: [1970] 286). Even though workers were employed from outside the local community, this does not mean, however, that we can recognise the existence of 'individual workers able to move around at their own free will.' Those areas outside the local community from which workers could be hired were more or less limited to a fixed number of specified locations. And, judging from the fact that there were also communities in the close vicinity of whaling groups where there is virtually no evidence of whaling related employment, it is thought that there must have been group employment, organised under the auspices of the local community, of workers with particular skills (Hidemura 1952b: 80). In regard to allowances, it is claimed that in Tosa from the first half of the nineteenth century, the move to paid labour started with the relatively unskilled 'barn' processing work, where there was a high proportion of day labourers and eventually also spread to those involved in whale catching activity, reaching the point, in the so-called 'enlightened Meiji era,' when all whaling workers, in theory at least, were stipulated as waged employees (Izukawa 1943: 208–14, 224–37, 420–32, 513–18 ([1973c]: 228–34, 244–57, 452–64, 545–50)). On the other hand, in northwest Kyūshū there is documentary evidence from the seventeenth century indicating that workers were paid a monetary wage. Further, since there were numerous whaling groups in this area,[7] to secure a labour force, wages were sometimes paid in advance as a kind of security deposit (Hidemura 1952b: 71, 91–104).

Let us now move on to look a little more closely at hereditary succession. In Taiji the *hazashi* remained a hereditary position throughout the entire history of net whaling. No matter how skilled they might be, the general whaling workers could never become *hazashi* (Kumano Taijiura hogeishi hensan iinkai 1969: 390–1, 524–30). On the other hand, it seems that in Tosa there was a system of selection by examination and those seeking to become *hazashi* could apply by submitting a 'letter of aspiration' (*nozomijō*) (Fukumoto 1978: 111–14; Izukawa 1943: 219 ([1973c]: 239)). In actual practice, however, it seems it was easier to be selected a *hazashi* if you had a close relative who had served in that position (Izukawa 1943: 426–8 ([1973c]: 458–60)). In this area, Ukitsu in Tosa, many of the *ogiri* ('fish cutters') engaged in processing activity had formerly been *hazashi* on the *mossōbune* (the boats that transport the whale after capture) and had been assigned to this work because they had been unable to move up to more prestigious employment as *hazashi* on the elite *sekobune* (pursuit boats) (Yoshioka 1938: 5–6, 46–8 ([1973]: 429–30, 470–2)).

So, as we have seen above, while it may be said that net whaling was closely incorporated into the local community, with workers arranged into a hierarchical pyramid of heredity and status that reflected the contemporary social structure (Morita 1994: 152), there was considerable variation from one whaling group to another and we can detect signs that it was on the road of change.

The introduction of Norwegian-style whaling

The introduction of whale-catching technology

In this section, in considering the actual introduction of Norwegian whaling technology, I shall be focusing on the accounts by Nihon En'yō Fisheries, the predecessor of Tōyō Whaling, of its own involvement in the process. First, let us look at the whale catching activity itself.

The Russian Count Keizerling Pacific Whaling Company established in 1894 (reorganised in 1899 into the Count Keizerling Pacific Whaling and Fisheries Company. Abbreviated below to Russian Pacific Whaling) and the Holme Ringer Company (Hōmu Ringā Shōkai) in Nagasaki, which acted as an agent for the 'Anglo-Russian Union,' an association consisting of two Russians and an Englishman who began Norwegian-style whaling in 1898, carried out whaling operations in the waters around the Korean Peninsula and exported whale meat to the port of Nagasaki. In response to this challenge, Nihon En' yō Gyogyō (Nihon En'yō Fisheries) was officially established in Yamaguchi prefecture in July 1899 around the persons of Yamada Tōsaku, Kawakita Kanshichi and Oka Jūrō, later to become president of Tōyō Whaling, with the backing of relatives and Diet members (Chōsenkai Tsūgyo Kumiai Rengōkai

1902a: 32–3; Emi 1907:121–4; Tōyō Hogei K.K., 1910: 192–201).[8] And in preparation for the beginning of actual whaling operations, in April of the same year, Oka was able to engage the services of a Norwegian harpoon gunner whose contract with Russian Pacific Whaling had run out and, with a little persuasion from the harpoon gunner, was able to employ an additional three Norwegian seamen. Later, in May 1899, at the general meeting of the company's backers, opinions were divided on the question of whether to acquire a Norwegian-style whaling vessel from abroad or build one in Japan, but eventually they come down in favour of the Japanese option on the grounds that waiting for a vessel to be made overseas would unnecessarily delay the beginning of operations. So, in June of that year, with the cooperation of the Norwegian harpoon gunner and crewmen, work commenced on the building of a whale catcher at the Ishikawajima dockyard in Tokyo. But it was decided to import whaling equipment, such as the harpoon gun, harpoons and harpoon ropes, directly from Norway through the Mitsui Trading Company (Mitsui Bussan). Meanwhile, at around the same time, Oka, who had been seconded part-time to the Ministry of Agriculture and Commerce, left on a trip to Europe and America in order to look into the Norwegian-style whaling industry. From May to December, first in Norway he observed and investigated the manufacture of whaling ships and equipment and actually witnessed operations at the whaling grounds. Then, on his way back to Japan, he toured the whaling industry facilities on the Atlantic coast of North America (Chōsenkai Tsūgyo Kumiai Rengōkai 1902a: 33(Emi 1907: 24–5; Tōyō Hogei K.K. 1910: 201–6).

Nihon En'yō Fisheries commenced actual operations with its new whale catcher, the *Daiichi Chōshū Maru*, in January 1900. In February of the same year, to counter whaling concessions Keizerling had acquired in 1899 in three locations on the Korean Peninsula, namely Ulsan in Gyeongsang Namdo, Jangjeon in Gangweondo and Mayangdo island in Hamgyeongdo, the company, in the name of its managing director Kawakita, gained monopoly whaling rights (the area of the concession unspecified) from the then Korean government.[9] But a stream of mechanical problems with the ship meant that the first and second seasons of whaling in the *Daiichi Chōshū Maru* were not particularly gratifying. As a result, direct operations were curtailed and it was determined to charter the whale catcher *Olga* for eight months from the ship's owners, the Holme Ringer Company, who had decided in the summer of 1901 to either sell or lease the vessel. But in December 1901 the

Daiichi Chōshū Maru after leaving Jangjeon bound for the waters off Wonsan hit a reef and sank. Subsequent attempts to raise the vessel ended in failure. Nihon En'yō Fisheries had used ninety per cent of its ¥100,000 capital in the exercise and was on the verge of liquidation, but it managed to overcome the crisis by renegotiating its contract with Holme Ringer for the lease of the *Olga* and at the same time chartering the whaling ships *Rex* and *Regina* from the Norwegian company Rex. This change in policy from using its own ships to relying on chartered vessels proved successful and Nihon En'yō Fisheries was able to turn around its fortunes (Chōsenkai Tsūgyo Kumiai Rengōkai 1902a: 33–4; Emi 1907: 125–41; Tōyō Hogei K.K. 1910: 191–2, 206–28. See also Chapter 5, note 5). From then until 1907–8 when Nagasaki Hogei Gōshi Kaisha, Dai Nippon Hogei Kabushiki Kaisha (Dai Nippon Whaling) and Teikoku Suisan (Teikoku Fisheries) started to build whale catchers domestically, Japanese whaling companies, with the exception, as we shall see later, of those that had captured Russian whaling ships, obtained their whale catchers either by chartering Norwegian-made ships or having them made to order in Norwegian shipyards (Tōyō Hogei K.K. 1910: 251–9).[10]

When relations between Japan and Russia were suspended with the outbreak of the Russo-Japanese war in February 1904, four ships of the Russian Pacific Whaling Company that were either anchored in Japanese ports or cruising in the waters around the Korean peninsula were confiscated by the Japanese government. Three of these, the catcher boat, *Nikolai*, the flensing ship, *Mikhael* and the transport ship, *Lesny* were officially 'captured.' The Nikkan Hogei Kabushiki Kaisha (Nikkan Whaling) was hurriedly formed under the patronage of fourteen members of the Diet in the hope of winning the right to use these vessels, but this company eventually merged with Nihon En'yō Fisheries, which was also petitioning for the captured ships, to form Tōyō Fisheries in September 1904. Also, in January 1904, Oka had succeeded, on behalf of Nihon En'yō Fisheries, in winning virtually the same whaling rights that had been afforded to Keizerling, including the three concessions at Ulsan, Jangjeon and Mayang Island. It later came to Oka's attention that Russian Pacific Whaling, which had stopped whaling because of the war and the loss of its ships, had fallen more than a year behind in its payment of the whaling concession tax and was hence in breach of the conditions of its contract. He urged the Korean government of the day to enforce the terms of the contract and confiscate all land, dwellings, commercial buildings and equipment within the concessions. Oka then succeeded, on the 1st May 1905, in persuading the Koreans to reassign the repossessed land and infrastructure to him. From this point in time until the end of World War II, whaling rights around the Korean peninsula were entirely in the hands of Japanese whaling companies (Nōrinshō Suisankyoku 1939: 5–7; Park 1995: 217–8, 278–93, 311–22; Tōyō Hogei K.K. 1910: 230–9).

Later, in 1906, Tōyō Whaling established whaling stations in a number of locations within Japan; not only those areas such as Kōchi and Wakayama where they practised net whaling, but also in places like Chōshi in Chiba and Ayukawa in Miyagi where there was no history of net whaling (Tōyō Hogei K.K. 1910: 242–7). Moreover, even on the Korean peninsula where whaling operations were commenced, we need to be aware of the fact that there seems to have been no whaling practised, certainly not on the scale of Japanese net whaling. As evidence we have the following report from someone who spent approximately one year in Pusan.

> At the time, during my voyages, whenever I saw such large numbers of whales, I was invariably both surprised and baffled by the fact that although our country and that land are not separated by any great distance, it is only there that we find such profusion of whales. After landing in Pusan, I asked one of the natives of that place why this should be so and he answered me thus. 'In Chōsen (Korea) since ancient times, whales, because they drive the fish towards us, have been worshipped as gods of the sea and hence we would never think of catching them.' It seems that in our country too, in former times whales were also considered sea gods and the general custom decreed that they should not be taken. Anyway, perhaps this is one reason why the whales are so numerous (Kaneki 1883: 11–12).

Further, after the Russo-Japanese war, the number of whaling companies mushroomed, spurred on by the success of Tōyō Whaling (See Figure 1.1). Among those to appear on the scene at this time were some, like Daitō Gyogyō (Daitō Fisheries) (1907) and Tosa Hogei Gōmeigaisha (Tosa Whaling Unlimited Partnership) (1907) from Kōchi, Kii Suisan Kabushiki Kaisha (Kii Fisheries) (1907) from Wakayama and Nagato Hogei Kabushiki Kaisha (Nagato Whaling) from Yamaguchi, that included locals who had previously been directly involved in the now defunct net whaling industry.[11] But these too were eventually absorbed into the established whaling companies like Tōyō Whaling and Hayashikane Shōten (Tōyō Hogei K.K. 1910: 251–68).

Let us now consider the composition of whaling crews and the actual nature of whale catching activities at this time.

Tables 1.3 and 1.4 illustrate respectively the composition of the crews of whale catcher boats as they must have been in the period 1899 to 1900 and the situation in 1901 when whaling companies were operating off the coasts of the Korean peninsula. Further, Table 1.5 has been produced from the description of Emi Suiin, who observed Tōyō Fisheries' whaling operations in Ulsan in 1906 and reflects the composition of crews in the whale catcher boats at that time. First, one point that immediately emerges clearly from these figures is that the harpoon gunners, who occupied a very important position and had high status, as we can see from the scale of wages in Table 1.1, were all Norwegians. Secondly, we see that those higher ranking crew members like the captain and engineer tend to be of the same nationality as the managers of the whaling company.

The third point to note is that the general seamen tended to come from the area in which the whaling company based its operations (Koreans in the case of Russian Pacific Whaling and Nihon En'yō Fisheries and Japanese for the Holme Ringer Company) or from a completely unrelated country (e.g. the Chinese employed by Russian Pacific Whaling and Holme Ringer Company). And eventually, as in the case of Tōyō Fisheries' use of the whale catcher *Rex*, we see the emergence of vessels managed by persons of different nationality from the managers of the whaling company.

Here let me add a note on the characteristics of whale catching activity at this time as we can piece it together from Emi's record.

A characteristic of Norwegian-style whaling was that there was no on-shore lookout (*yamami*). Instead, whales were spotted from a crow's nest attached to the mast of the ship. Moreover, as the record mentions that 'sailors took it in turns to climb up to the crow's nest to look out for surfacing whales' we can assume that the special status that had been accorded to the *yamami*, as director of the whaling operation, or, as in the Taiji whaling group where a member of the most influential family performed the role of *yamami* (Kumano Taijiura hogeishi hensan iinkai 1969: 388–90), had disappeared (Emi 1907: 19). A further characteristic is the way in which three men in a small boat are sent, after the whale has been hit by the bolt from the whaling gun, to deal the death blow with handheld harpoons (Emi 1907: 52–3, 104–5).

Let me now summarise the points made above regarding the process of introduction of whaling technology. First, is the fact that the introduction of Norwegian-style whaling technology ultimately

Table 1.3: Composition of crews in whale catchers working off the Korean Peninsula (1899–1900)

Pacific Whaling and Fisheries Company of Count H. H. Keizerling

Ship	*Nikolai*		*Georgii*	
Registry	Russia (company owned)		Russia (company owned)	
Power	Steam		Steam	
	Nationality/Name	**Number**	**Nationality/Name**	**Number**
Captain	Russian/Kapustin	1	Russian/Shabalin	1
Harpoon gunner / Steersman	Norwegian	1	Norwegian	1
Engineer	Russian	1	Russian	1
Seamen / Stoker	Korean	8	Korean	8
Steward / Ship's boy	Chinese	2	Chinese	2
Total		13		13

Nihon En'yō Fisheries

Ship	*Daiichi Chōshū Maru*	
Registry	Japan (company owned)	
Power	Steam	
	Nationality/Name	**Number**
Captain	Japanese?	1
Harpoon gunner	Norwegian	1
Steersman	Japanese?	1
Engineer	Japanese?	1
Seamen and stokers	Korean	2
	Japanese?	11
Steward	–	–
Ship's boy	–	–
Total		17

Notes:

1. The occupations and the nationality of the crew, and the spellings of ships' names and personal names follow the Chōsen Fisheries Association (Chōsen Gyogyō Kyōkai 1900). – indicates that no information has been provided.

2. The nationality of several members of the crew of the Daiichi Chōshū Maru is not clearly indicated, but given that Norwegians and Koreans are differentiated, it seems likely that these are Japanese.

3. According to the Chōsen Fisheries Association, in addition to the above companies, the Holme-Ringer Company, En'yō Whaling and the Sanshū Whaling Group (net whaling from Kagawa prefecture) were also active at this time in whaling in coastal waters around the Korean Peninsula, but the details of their organisation are unclear (Chōsen Fisheries Association 1900: 16–17).

Sources: Chōsen Gyogyō Kyōkai (1900), 'Kankai hogeigyō no ippan,' *Dai Nihon suisankaihō*, 212, pp. 4–19; 'Nichiro ryōkokujin no Kankai hogei jōkyō' (1904), *Dai Nihon suisankaihō*, 260, pp. 34–36; Tōyō Hogei K.K. ed. (1910), *Honpō no Noeruēshiki hogeishi*: 203–5.

Table 1.4:   Composition of crews in whale catchers working off the Korean Peninsula (1901)

Pacific Whaling and Fisheries Company of Count H. H. Keizerling

Ship	*Nikolai*		*Georgii*	
Registry	Russia (Company owned)		Russia (Company owned)	
Power	Steam		Steam	

	Nationality/Name	Number	Nationality/Name	Number
Captain	Russian/Kapustin	1	Russian/Tiderman	1
Harpoon gunner / Steersman	Norwegian/Amundsen	1	Norwegian/Melsom	1
Apprentice gunner	–	–	–	–
Chief engineer	–	–	–	–
Engineer	Russian	1	Russian	1
Seaman / Stoker	Korean	8	Korean	8
Ships' boy	Chinese	1	Chinese	1
Total		**12**		**12**

Nihon En'yō Fisheries

Ship	*Daiichi Chōshū Maru*	
Registry	Japan (Company owned)	
Power	Steam	

	Nationality/Name	Number
Captain	Japanese/ Hamano Tōtarō	1
Harpoon gunner	Norwegian/ Morten Pedersen	1
Steersman	?	1
Apprentice gunner	Japanese/ Nomura Sadatsugu	1
Chief engineer	?	1
Engineer	–	–
Seamen	?	6
Stoker	?	4
Ship's boy	?	2
Total		**17**

Table 1.4: continued

Holme-Ringer Company

Ship	Olga	
Registry	Russia or Norway (Company owned)	
Power	Steam	

	Nationality/Name	Number
Captain	Russian	1
Harpoon gunner	Norwegian/ Olsen	1
Steersman	British	1
Apprentice gunner	–	1
Chief engineer	?	1
Engineer	British	1
Seamen Stoker	} Japanese	8
Ship's boy	Chinese	2
Total		14

Notes:

1. The occupations and the nationality of the crew, and the spellings of ships' names and personal names follow the Federation of Fishing Unions working in the Korea Sea (*Chōsenkai Tsūgyo Kumiai Rengōkai* 1902a and 1902b). – indicates that no information has been provided.

2. The term 'operating vessel' in the source materials is taken to refer to whale catcher boats.

3. The source materials put the total crew of the *Nikolai* and the *Georgii* at thirteen, but this is presumably an error. Here corrected to twelve.

4. The nationality of several members of the crew of the *Daiichi Chōshū Maru* is not given in the sources. However, when the vessel sank in 1901, to judge from the names of the victims, one Korean and two Japanese were drowned. Here I have not speculated on the nationality of these crewmen as I did in Table 1.3.

5. The registry of the Olga is given as Norway in '*Nichiro ryōkokujin no Kankai hogei jōkyō*' and as Russia in '*Chōsenkai hogeigyō*.'

Sources: Chōsenkai Tsūgyo Kumiai Rengōkai (1902a), 'Chōsenkai hogeigyō', *Dai Nihon suisankaihō*, 234, pp. 24–37; Chōsenkai Tsūgyo Kumiai Rengōkai (1902b), 'Chōsenkai hogeigyō', *Dai Nihon suisankaihō*, 235, pp. 21–37; 'Nichiro ryōkokujin no Kankai hogei jōkyō' (1904), *Dai Nihon suisankaihō*, 260, pp. 34–36; Tōyō Hogei K.K. ed. (1910), *Honpō no Noeruēshiki hogeishi*, 191–2, 203–5, 216–19.

took the form of using ships and equipment manufactured in Norway and also employing Norwegian nationals as harpoon gunners who were the bearers of this technology. Following on from the success of the experiment with imported technology, whaling

Table 1.5: Whale catcher boats of Tōyō Gyogyō K. K. (1906)

	Nikolai		Rex	
	Nationality/Name	Number	Nationality/Name	Number
Captain	Japanese/Natsume Ichitarō	1	Norwegian/Melson (captain and gunner)	1
Harpoon gunner	Norwegian/Jordensen	1		
Chief engineer	Japanese/Yamaguchi	1	Norwegian/Magaisen	1
Deck				
Boatswain	Japanese?	1	Korean/Kim Jin Hee	1
Seamen	Japanese/Ōoka, Kogano (and two others)	4	Norwegian/Hans (seaman)	1
Engine room (stokers)	?	5		
Galley			Korean (and other general seamen)	9
Cook	?	1		
Boy	1 Japanese	2		
Total		16		13

Notes:

1. The occupations of the crew and the spellings of ships' names and personal names follow the source cited below.

2. As the boatswain and one of the boys on the *Nikolai* communicated with the author (Emi) in Japanese, it is assumed they were Japanese.

3. The Norwegian seamen, Hans, was apparently the son of a deceased friend of the gunner Melson. He was probably still a boy and had joined the crew of the *Rex* to start his training as a harpoon gunner.

4. In addition the source mentions the chaser, *Oruga Maru*, but there is too little information regarding the composition of its crew for it to be included in this table.

Source: Emi, Suiin (1907), *Jitchi tanken: hogeisen*, Hakubunkan.

boats soon came to be built in earnest in Japan too and eventually, as we shall see below, Japanese began to be employed as harpoon gunners. It is clear that in this process, it was not the managers of net whaling concerns who were responsible for the introduction of Norwegian-style whaling technology, neither was it the former *hazashi* harpooners who became harpoon gunners. Moreover, the fact that whaling crews came to be composed of people of various nationalities, clearly suggests that the hereditary, class-based social structure of net whaling was being replaced by a new social structure sited on the whaling company.

At this point we move on to consider the composition of the whaling labour force as we enter stage three in our chronology. The materials available for our investigation, however, are very

Table 1.6: The crew of the shipwrecked Inazuma Maru (1933)

Captain and Harpoon gunner	Okayama Tomihide
Chief engineer	Kagawa Kaname
Steersman	Yoshikawa Kamesuke
Boatswain	Murazume Iwakichi
Seaman	Kim Jang Uk (or Chinese Jin Zhang Xu?) Ogawa Yoshimi Yasunaga Naokichi Rachi Toraji
Chief stoker	Nishikawa Tomoya
Oiler	Kamijima Katsuo
Stoker	Kim Weon Jong (or Chinese Jin Yuan Zhong ?) Kim U Geun (or Chinese Jin You Gen ?)
Steward	Liu Zhao Meng (Chinese?) or Yu Jo Mang (Korean ?)
Total	13

Source: Park, Koo-Byong (1995), *Jeungbopan Hanbando eonhae pogyeongsa*, Pusan: Toseochulpan Mincheok Munhwa, 287.

limited indeed. One such source is Table 1.6. This is taken from a commemorative plaque bearing the names of the victims of the tragedy of the Tōyō Whaling whale catcher *Inazuma Maru* that vanished with all hands after leaving the port of Seogwipo on Jeju Island. From the plaque we can see that by this time a Japanese held the position of harpoon gunner and that Koreans were counted among the seamen.[12] According to Park Koo-Byong, before the end of World War II the highest position to which a Korean seaman could aspire was that of boatswain, so we can assume that Koreans could not rise to become harpoon gunners, those with the highest status among the crew (Park 1995: 324, 331).

The introduction of processing technology

Let us now consider the processing technology. When Nihon En'yō Fisheries started its operation in January 1900, the flensing ship *Chiyo Maru* sailed in tandem with the *Daiichi Chōshū Maru* and in 1901 when the company chartered the *Olga*, it also leased the flensing ship *Kōsei Maru*. One reason for the introduction of flensing ships at this time was probably because it was not possible to carry out processing operations on land, as the company had been unable to establish whaling stations on the Korean peninsula (Tōyō Hogei K.K. 1910: 209–16). But judging from the fact that, in 1909, when the company had managed to obtain twenty whaling stations, it continued to run thirteen flensing ships, we can assume that flensing ships had come to play an established role as new technology. That is to say, mobile flensing ships were considered efficient when exploring new whaling grounds or for temporary operations, as they made it possible to carry out processing operations without establishing permanent whaling stations on shore (Kondō 2001: 220–1).

Now, as I mentioned before, among the vessels captured during the Russo-Japanese war and lent to Tōyō Fisheries was the flensing ship *Mikhael* (*Mihairu*). This had a net registered tonnage of 2,144 tons (by comparison, the *Chiyo Maru* had a gross tonnage of 154 tons) and a crew of 'thirty Russian and other Europeans, ten Japanese, over fifty Chinese and a dozen or so Koreans' (*Nichiro ryōkokujin no Hankai hogei jōkyō*, 1904: 35–6), distributed over five levels of decks, each engaged in various operations such as oil extraction or the pulverising of whale bone with the aid of twenty machines. It can be seen in every sense as a precursor of the modern whaling factory ship (*Nichiro ryōkokujin no Hankai hogei jōkyō*, 1904: 35–6; Tōyō Hogei K.K. 1910: 187–8, 209). Then, as I pointed out above, Tōyō Fisheries, after confiscating all Russian Pacific Whaling's land, dwellings, commercial buildings and equipment, proceeded to take over the whaling concessions that company had held. We can consider, therefore, that Tōyō Fisheries had been able to acquire new processing technology through its windfall gains from the Russo-Japanese war[13] and that this very much set the tone for the process of Japanese encroachment on the Korean peninsula.

At this point I should like to take a slightly more detailed look at the nature of the work and the composition of the workforce in processing operations. Tables 1.7 and 1.8 show the composition of crews on the flensing ships for each whaling company for the same periods, respectively, as Tables 1.3 and 1.4. It is clear from these charts that, firstly, those employed as the leaders in Nihon En'yō Fisheries' processing activity were a Norwegian and another 'very experienced individual with in excess of two years experience working as chief flenser on the Russian ship *Olga*.' Secondly, we notice that companies not under Japanese management, like Russian Pacific Whaling and the Holme Ringer Company employ Japanese as general process workers. Of these, the five employees in the salting room shown in figure 1.12 were 'Japanese from the Gotō Archipelago (off northwest Kyūshū)' (Chōsen Gyogyō Kyōkai 1900: 13) who had been employed since 1900 to salt for sale in Nagasaki, whale meat that had until then been 'sold to Koreans at throw away prices' (Chōsen Gyogyō Kyōkai 1900: 12–15). However, even when the transport of the harvested whale took a considerable time, the whale meat produced by Nihon En'yō Fisheries by the technique, in which the *hazashi* slices the flesh from between the ribs to remove the blood from the carcass, could be sold off at a high price ('the going rate held up well at around fourteen or fifteen sen a catty (600 grams)' (Chōsenkai Tsūgyo Kumiai Rengōkai 1902b: 34). In contrast, the whale meat produced by Russian Pacific Whaling, which had not been put through any process of this kind, was sold at a very low price ('a catty would sell for barely two or three sen' (Chōsenkai Tsūgyo Kumiai Rengōkai 1902b: 34).[14] The reasons given to explain why Russian Pacific Whaling did not employ this method were, that the whale carcass with the blood removed tended to sink and make transport exceedingly difficult, and that 'foreign whalers generally use the blubber and the whalebone and hardly give a second thought to the meat' (Chōsenkai Tsūgyo Kumiai Rengōkai 1902b: 34, 33–6).

Further, it is interesting to note here that the term '*hazashi*,' a role that had disappeared with the advent of Norwegian-style whale catching activity, resurfaces as the title for the man who performs an important task in the processing operation.[15]

For a description of the actual processing activity at this time we now turn to the account provided by Emi. There is a support vessel moored alongside the wharf of the whaling station. The flensing ship *Chiyo Maru*, which has taken delivery of the whale from the catcher boat, is drawn up gunwale to gunwale beside the support vessel. First the flensers approach the *Chiyo Maru* in a barge and pass a hook attached to the end of a wire cable through the pectoral fin of the whale positioned against the hull of the *Chiyo Maru*. As the winch housed on the *Chiyo Maru* starts to reel in the wire cable, the flensers immediately insert a cut with the large flensing spades (*ōbōchō*) so that the section around the pectoral fin is severed from the carcass and raised to the deck of the ship. On the flensing ship, winches are used in the same way in the repeated skinning and slicing operations. At the same time the butchers (*saikatsufu*) cut the lumps of meat hanging above the deck into sections of roughly the same size. The large lumps of meat are carried to the deck of the support ship by the *kagihiki* (pullers) where they are cut into

Table 1.7: Composition of crews in flensing ships working off the Korean Peninsula (1899–1900)

Pacific Whaling and Fisheries Company of Count H. H. Keizerling

Ship	*Sibir'*		*Taiyō Maru*	
Registry	? (Company owned)		? (Vessel leased from the Japanese)	
Power	Steam and sail		Sail	
	Nationality/Name	**Number**	**Nationality/Name**	**Number**
Captain	Russian/Ivanov	1	Japanese/ Yoshida Masutarō	1
Steersman	Russian	1		
Engineer	Russian	1		
Seaman and stokers	Korean	4		
Steward	Chinese	2		
Ship's boy				
Supervisor	–	–		
Accountant	–	–		
Flenser-in-charge	Russian	1	Japanese	6
Foreman	–	–		
Flenser	–	–		
Salter	Japanese	5		
Blacksmith	–	–		
Carpenter	–	–		
Other	–	–		
Total		15		7

Pacific Whaling and Fisheries Company of Count H. H. Keizerling

Ship	*Kameran*	
Registry	?(company owned)	
Power	Sail	
	Nationality/Name	**Number**
Captain	–	–
Steersman	–	–
Engineer	–	–
Seaman	Japanese	2 (including steward)
Stokers	–	–
Steward	Japanese	2
Ship's boy	–	–
Supivisor	–	–
Accountant	–	–
Flenser-in-charge	–	–
Foreman	Russian	1
Flenser	Chinese	20

Table 1.7: continued

	Nationality/Name	Number
Salter	–	–
Blacksmith	Russian	1
Carpenter	–	–
Other	Deputy manager/ Hugo Keizerling	2
	Wonsan customs officer/ Mutsu Kojirō	
Total		26

Nihon En'yō Fisheries

Ship	*Chiyo Maru*
Registry	Japan (Company owned)
Power	Sail

	Nationality/Name	Number
Captain	Japanese?	1
Steersman	–	–
Engineer	–	–
Seamen	Japanese?	10
Stoker	–	–
Steward	–	–
Ship's boy	–	–
Supervisor	Japanese?	1
Accountant	Japanese?	1
Flenser-in-charge	Norwegian	1
Foreman	–	–
Flenser	Japanese?	8
Salter	–	–
Blacksmith	Japanese?	1
Carpenter	Japanese?	1
Other	–	–
Total		24

Notes:
1. The occupations and the nationality of the crew, and the spellings of ships' names and personal names follow the Chōsen Fisheries Association (Chōsen Gyogyō Kyōkai 1900). '–' indicates that no information has been provided.
2. The nationality of several members of the crew of the *Chiyo Maru* is not specified, but given that Norwegians and Koreans are distinguished in the case of the Daiichi Chōshū Maru, it is assumed they are Japanese.
3. According to the Chōsen Fisheries Association, in addition to the above companies, the Holme-Ringer Company, En'yō Whaling and the Sanshū Whaling Group (net whaling from Kagawa prefecture) were also active at this time in whaling in coastal waters around the Korean Peninsula, but the details of their organisation are unclear (Chōsen Fisheries Association 1900: 16–17).

Table 1.7: continued

4. I take the term *saikaisen* (literally, 'butchering ship') in the source materials to be equivalent to *kaibōsen* (flensing ship). The *Kameran* is listed as an 'executive staff ship,' but since it carried flensers on board I have treated it as a flensing ship.

Sources: Chōsen Gyogyō Kyōkai (1900), 'Kankai hogeigyō no ippan,' *Dai Nihon suisankaihō*, 212, pp. 4–19; 'Nichiro ryōkokujin no Kankai hogei jōkyō' (1904), *Dai Nihon suisankaihō*, 260, pp. 34–36.

the various smaller pieces by an experienced butcher using a small knife (*kobōchō*). Finally, these smaller pieces are again carried by the *kagihiki* along the wharf to the freezer room (Emi 1907: 58–67). The numbers of personnel involved in these processing operations were as follows; three flensers on the barge, eight butchers on the flensing ship, five or six *kagihiki* runners transporting meat between the flensing ship and the support vessel and ten butchers on the support vessel. The nationality of these workers is not recorded. But there is a note to the effect that most of the dozen or so *kagihiki* who took the whale meat from the support vessel to the freezing room were Koreans. Moreover, in addition to these workers, Emi records the names of the two supervisors, Yoshizu Gonzaburō and Okata Rihei and the assistant supervisors, Matsuya and Kameya (Emi 1907: 14, 62, 66–7).

At this time it is generally accepted that dissection of the whale carcass was carried out by using either flensing ships, or later, with the establishment of on-shore whaling stations, the bok (pile) wharf technique (Kondō 2001: 220–1, 230, 242–3; Tōyō Hogei K.K. 1910: 114). The progress of the processing activity described by Emi suggests that he observed the former. For comparison, we have the following brief description of the bok wharf method. Two large piles are erected on the wharf. A crossbar is fixed near the top, joining the two piles, and two pulleys are attached to the crossbar. A wire cable attached to a winch is run over each of the pulleys and a hook is attached to the end of the cable. When a whale catcher comes alongside the wharf, a rope is put through the tail flukes of the whale hanging from the gunwale of the ship. A hook is passed around the rope and the whale is pulled towards the shore. At the same time a chain is wound around the tail section of the whale and then the hook is disengaged from the rope and hooked into the chain to raise the whale's body about one quarter of the way out of the water. At this point, the *kaibōshu* (flensers) (Tōyō Hogei K.K. 1910: 116) working from a *denmasen* (barge) (Tōyō Hogei K.K. 1910: 115) first make a cut around the carcass from the area of the genital

organs and this severed section is raised with a winch to hang above
the wharf. There it is cut into smaller sections by the *saikatsushu*
(butchers) (Tōyō Hogei K.K. 1910: 116). When this process is
finished, the remaining section of the carcass, which has been left
lying in the water alongside and parallel to the wharf, is winched up
in the same way so the skin can be removed and the flesh cut away,
after which the butchers repeat the process of dividing it up into
smaller specific cuts of meat (Kondō 2001: 220–2, 233; Tōyō Hogei
K.K. 1910: 114–17). Further, in 1907, in the Ayukawa station of
Tōyō Fisheries a new dissection technique was developed in which
the entire whale was pulled completely onto the shore. Henceforth,
this became the generally followed method and remains so to this
day (Kondō 2001: 242–3). There was, besides these techniques, one
employed by the Vancouver Island Whaling Company in the then
British colony of Canada in which 'the order of the processes differs
from that employed in Japan,' (Tōyō Hogei K.K. 1910: 158), in that
the whale is winched ashore by a chain wound around the base of the
tail (Abe 1908b: 8), then, first, as in Emi's description, the pectoral
fins are removed. Next, an incision is made from the whale's head
to its tail and the skin is removed with a winch (Abe 1908b: 8–9;
Tōyō Hogei K.K. 1910: 157–60).[16] Even so, the method itself of using
a winch to flay the skin, including Russian Pacific Whaling's use
of the technique (Chōsen Gyogyō Kyōkai 1900: 12–13; Chōsenkai
Tsūgyo Kumiai Rengōkai 1902b: 33–4), [17] should probably be seen
as an extension of Norwegian-style processing technology.

Further, in comparing these with the processing methods of net
whaling we can identify a number of points of difference. We have
numerous descriptions of the whale being pulled onto the beach
head first[18] and, since a reel was used rather than a winch, large
numbers of workers, including day labourers, were required.[19]

Let me now summarise what we have affirmed regarding the
introduction of processing technology. First, it is clear that the
Norwegian-style processing was not simply a continuation of the
'barn' processing activity associated with net whaling. Because it
was combined with Norwegian-style whale catching, it necessitated
the introduction of new technology in the form of flensing vessels
and winches. In addition, it was Norwegians and others who had
gained experience in whaling companies run by non-Japanese
who shouldered the introduction of the technology. Further, we
can consider that most of the new processing technology had
been introduced through the confiscation of flensing ships and
processing plants on the Korean peninsula as a result of the Russo-

Table 1.8: Composition of crews in flensing ships working off the Korean Peninsula (1901)

Pacific Whaling and Fisheries Company of Count H. H. Keizerling

Ship	*Gorlitsa*		*Lesnik*	
Registry	Russia (company owned)		Russia (company owned)	
Power	Sail		Sail	

	Nationality/Name	Number	Nationality/Name	Number
Captain	Russian/ Volsov	1	Japanese/ Yoshida Masutarō	1
Steersman	Russian	1	Japanese	1
Seaman Supervisor Chief flenser	Russians, Chinese and Koreans	6		
Flenser-in-charge	Russian	1	Russian and Korean	6
Flenser	Chinese	20		
Salter Blacksmith Carpenter	Russians, Chinese and Koreans	6		
Other			Wonsan customs officer, Riku (*sic*) Kojirō	1
Total		29		9

Nihon En'yō Fisheries

Ship	*Chiyo Maru*		*Sumiyoshi Maru*	
Registry	Japan (company owned)		?	
Power	Sail		Sail	

	Nationality/Name	Number	Nationality/Name	Number
Captain	Japanese/ Aida Eikichi	1 (also chief flenser)	Japanese/ Shionoya Matagorō	1
Steersman	–	–	–	–
Seaman	?	8	?	2
Supervisor	Suko Kamematsu	1	–	–
Chief flenser	–	(1)	–	–
Flenser-in-charge	–	–	–	–
Flenser	–	–	?	18
Salter	–	–	–	–
Blacksmith	?	1	–	–
Carpenter	?	1	–	–
Other	Wonsan customs officer/ Suzuki Itsuji	1	–	–
Total		13		21

Table 1.8: continued

Holme-Ringer Company

Ship	*Kōsei Maru*
Registry	Japan
Power	Sail

	Nationality/Name	Number
Captain	Japanese/Mikami Torajirō	1
Steersman	–	–
Seaman	Japanese	8
Supervisor–	–	–
Chief flenser–	–	–
Flenser-in-charge	–	–
Flenser	Japanese	8 (combined flensers and salters)
Salter		
Blacksmith	–	–
Carpenter	–	–
Other	–	–
Total		17

Notes:

1. The occupations and the nationality of the crew, and the spellings of ships' names and personal names follow the Federation of Fishing Unions working in the Korea Sea (Chōsenkai Tsūgyo Kumiai Rengōkai 1902a and 1902b). '–' indicates that no information has been provided.

2. I take the term *saikaisen* (literally, 'butchering ship') in the source materials to be equivalent to *kaibōsen* (flensing ship). The *Sumiyoshi Maru* is listed as a 'storage ship' (chozōsen), but since it carried flensers on board I have treated it as a flensing ship.

3. The nationality of several members of the crew of the *Chiyo Maru* and the *Sumiyoshi Maru* is not provided. Here, as in Table 1.9, I have left these cases unresolved.

Sources: Chōsenkai Tsūgyo Kumiai Rengōkai (1902a), 'Chōsenkai hogeigyō,' *Dai Nihon suisankaihō*, 234, pp. 24–37; Chōsenkai Tsūgyo Kumiai Rengōkai (1902b), 'Chōsenkai hogeigyō,' *Dai Nihon suisankaihō*, 235, pp. 21–37; 'Nichiro ryōkokujin no Kankai hogei jōkyō' (1904), *Dai Nihon suisankaihō*, 260, pp. 34–36.

Japanese war. In addition, it has come to light that 'Japanese' were employed as general process workers in companies managed by non-Japanese and that it was these people who produced whale meat for consumption. Further, we saw how Koreans were employed as general process workers in Japanese whaling companies.

Coming into the third period, again we have very little data on processing activity and the composition of the workforce, as indeed is also the case with whale catching activity *per se*. One of these

sources is a record of Koreans working at 'home-island' (*naichi*)
whaling stations. This can be found in a survey on *Chōsenjin
rōdōsha ni kansuru jōkyō* (On the situation of Korean workers)
carried out in 1924 by the Bureau of Social Affairs, Ministry of the
Interior. In the 'supplementary reference tables' appended to the
report, the type of employment column for Iwate prefecture carries
the entry 'fisheries (miscellaneous tasks in whaling company)'
(Naimushō Shakaikyoku Daiichibu 1924: 1975: 450–540). It is
assumed that the whaling company mentioned here is Tōyō Whaling
which at the time had a station in Kamaishi. But this source gives
us no indication of how these workers had come all the way from
the Korean peninsula to work in Iwate.[20] But at least we can say that
at this time whaling companies needed Korean workers to perform
'miscellaneous tasks,' in favour of workers from neighbouring areas
or those former net whaling workers who had been put out of work
with the advent of the more personnel-efficient Norwegian-style
whaling methods.[21]

The introduction of factory-ship whaling

Let us now consider the introduction of technology associated with
factory-ship whaling, focusing on the use of the factory vessel as a
mothership. First, in 1929 Tōyō Whaling tried to gain knowledge
of the new technology by having one of its staff (exactly who is
not clear) join the crew of a Norwegian factory ship, but their
request was refused and the plan was abandoned. Then in 1930, the
company purchased a passenger and freight vessel from Britain for
conversion to a factory ship and sailed the ship back to Japan, but
due to Tōyō Whaling's management slump mentioned above, this
venture also ended in failure (Baba 1942: 90–1). After this, Nihon
Sangyō Kabushiki Kaisha (Nihon Production Co.), with any eye
on the whaling industry, sought not only to gain the necessary
personnel and infrastructure by buying up whaling companies,
but planned to amalgamate all the whaling companies under the
company's banner. But as soon as it became apparent that this
amalgamation could not be achieved quickly, the company decided
to change its tactics and first purchase the largest whaling company,
Tōyō Whaling, resulting in the birth of Nihon Hogei Kabushiki
Kaisha (Nihon Whaling) in 1934. In June of that year the new
company succeeded in purchasing the Norwegian factory ship
Antarctic together with its associated fleet of five catcher boats.
Subsequently the ship was transferred to Japanese registration and

its name changed to *Antākuchikku Maru*. Crew members required for whale catching and processing activities were sent from Japan and on the voyage home they decided to carry out whaling trials in the Antarctic Ocean between Cape Town and Australia (Baba 1942: 92–101).

In the following year, 1935, the *Antākuchikku Maru* had its name changed for the third time to *Tonan Maru* and again set out to harvest whales in the Antarctic.

To counter this activity of Nihon Whaling, in 1935 Hayashikane Shōten initiated the building of the whaling factory ship *Nisshin Maru* in the Kawasaki Shipyards in Kobe. Further, in 1936 it established Taiyō Hogei and began participating in factory-ship whaling in the Antarctic Ocean. From this time onwards factory ships were built in Japanese dockyards, for example the *Daini Tonan Maru* (launched in 1937) and *Daisan Tonan Maru* (launched in 1938) of Nihon Suisan (Nihon Fisheries), a company formed in the wake of a restructuring of Nihon Whaling, Taiyō Hogei's *Daini Nisshin Maru* (launched in 1938) and the *Kyokuyō Maru* (launched in 1938) of Kyokuyō Hogei, which had been established in 1937 by Sumatora Takushoku Kabushiki Kaisha (Sumatra Colonial Co.) when it entered the whaling industry and which had begun factory-ship whaling in the Antarctic Ocean in 1938 (Baba 1942: 101–8, 263; Nōrinshō Suisankyoku 1939: 2).

What then, was the nature of the technology for harvesting and processing whales associated with the advent of factory-ship whaling? The various whaling companies operated fleets of between five and ten catcher boats. The catcher boats themselves were large in comparison with those used in coastal whaling (coastal whale catchers had a gross tonnage of 100 to 120 tons, whereas the whale catchers employed with factory ships ranged from 220 to 390 tons). The fuel burned had changed from coal to heavy oil and the engines from steam to diesel (Baba 1942: 158–68; Nōrinshō Suisankyoku 1939: 17–19). In addition, the need to maintain frequent contact with the factory ship meant that whale catchers had to be fitted with powerful wireless communications equipment (Baba 1942: 168–70). In addition, the rear-loading (shell-type) harpoon gun replaced the muzzle loading gun (in which a bag of gun powder was put in through the mouth of the barrel, followed by a round felt disc, cotton wadding and rubber packing all packed down tightly with an oak ram rod) and manilla hemp rope came to be used for the harpoon rope, alongside the so-called tarred rope (made from Japanese hemp soaked in high quality refined wood-tar). It is said that the rear loading gun and the manilla hemp rope were introduced with the purchase of the *Antarctic* mentioned above, as they were part of the equipment on the ancillary whale catchers (Baba 1942: 190–3; Kondō 2001: 303–5, 369–70). In addition to these innovations, the technique of using a pump to fill the abdominal cavity with air was introduced in order to prevent the whale from sinking directly after harvesting. This procedure not only made it easier to tow the whale secured to the hull of the ship, but also allowed for the marking of the catch with flags or lights, so that the pursuit of whales could resume soon after a whale was taken (Baba 1942: 209–10). [22]

It goes without saying that the introduction of the factory vessel as a mothership was the most noteworthy innovation in processing technology. As we have seen above, at the time of the introduction of Norwegian-style whaling, flensing ships were used. There was no great change from this pattern in Japanese whaling, but moored factory ships were apparently used as a convenient temporary expedient in islands in the Antarctic Ocean where it was difficult to build whaling stations and when seasonal variation, or changes in the productivity of fishing grounds, made it necessary to shift processing activity from one port to another (Baba 1942: 198–200). The main difference between the early flensing ships or moored factory vessels and the floating factories was mainly a matter of size. The smallest, the *Tonan Maru*, for example, had a net tonnage of 9,866 tons, compared with the largest, the *Daini Tonan Maru* with a net tonnage of 19,425 tons. It should also be noted here that whereas the early factory ships carried out the processing with the whale alongside the hull of the ship, as I described above, the factory ships introduced in this later period were equipped with a slipway in the stern of the ship, in which part of the upper deck was cut away and a ramp, just like a child's slippery-dip, led from there to water level. The whale was winched tail-first up this slipway to the upper deck (Baba 1942: 68–73, 173–80, 200; Nōrinshō Suisankyoku 1939: frontispiece illustration, 17).

In the actual processing of the whale carcass, there were increases in the size, efficiency and degree of automation of the tools used. For example, there were greater numbers of fixed winches employed, mechanised saws, powered by steam or electricity, were used for cutting through the bones and the boilers were upgraded from press boilers (the so-called 'steam can') to Hartman and Kuwana boilers, equipped with a revolving drum.

But, as was the case with whale catching activity itself, we can assume that the actual processing operation was more or less the

Table 1.9: Products of Japanese whaling (1935–1941)

	1935	1936	1937	1938	1939	1940	1941
Antarctic Ocean							
Whale oil (tons)	2,006	7,358	26,089	64,044	80,629	90,167	104,138
Salted meat (tons)	28	264	270	1,212	2,831	8,382	13,536
Other (tons)	–	–	–	212	1,682	2,609	2,451
North Pacific							
Whale oil (tons)						4,588	3,986
Salted meat (tons)						1,486	3,655
Other (tons)						–	–
Coastal							
Whale oil (tons)	4,305	4,883	5,509	4,471	4,015	4,360	4,286
Whale meat (tons)	19,350	19,892	16,382	19,142	17,801	17,801	21,476
Other (tons)	4,072	3,737	7,368	2,951	2,275	2,275	4,897

Notes:

1 Figures are rounded to the nearest ton. '–' indicates a lack of data. Data for the Antarctic are listed under the year in which the fishing season ended.

2 The figure for whale meat listed under coastal whaling includes the statistics for whale meat plus those for the tail fluke and the ventral pleats (*oba-une*).

3 The Antarctic figure for salted meat for the years 1938–41 refers to meat and skin as listed in 'Dainijūyonji Nōrinshō tōkeihyō'.

Sources: Maeda, Keijirō and Teraoka Yoshirō (1952), *Hogei,* Nihon Hogei Kyōkai, 116; Nōrin Suisanshō Tōkei Jōhōbu: Nōrin Tōkei Kenkyūkai (1979), *Suisangyō ruinen tōkei dainikan, vol.2*, Nōrin Tōkei Kyōkai; Nōrin Daijin Kanbō Tōkeika (1926–30, 1932–36, 1938–42), Daiichi – daijūhachiji nōrinshō tōkeihyō; Nōrinshō Nōgyō Kairyōkyoku Tōkei Chōsabu (1949–51), *Dainijūyon – nijūrokuji Nōrinshō tōkeihyō*, Nōrin Tōkei Kyōkai.

same as it had been in the previous period (Baba 1942: 180–9, 210–14; Kondō 2001: 308–10, 383, 386–92). However, there was a difference in the proportions of product produced. At the time of the introduction of Norwegian-style whaling (the second period), it seems the emphasis had been on whale meat production, while whale oil was considered a by-product, but, in this period, the main focus was on whale oil production and whale meat was regarded as being of secondary importance (Tōyō Hogei K.K. 1910: 56–65, 119–20, 163–4, 253, 257, *passim*). Against the historical background of the outbreak of the China-Japan War in July 1937 and the sudden eruption of World War II in Europe in September 1939, whaling companies poured all their energies into the production of whale oil [23] which was an important substance for military purposes and

could be exported to Europe to earn foreign currency (Matsuura
1944: 84–8; Ōmura 1938: ; Baba 1942: 3–4, 299–300). Certainly,
in contemporary coastal whaling and in factory-ship whaling in
the North Pacific, whale meat continued to be the more important
product (Baba 1942: 115–19, 200–2, 223–4). Further, in response
to calls for 'the complete use of the whale carcass' when the
International Whaling Agreement was signed in London in 1937,[24]
and to meet the Japanese domestic need to augment shortages
in food and resources occasioned by the war with China, they
remodelled factory ships originally dedicated to specialise in
whale oil production, and, particularly after the introduction of
refrigerated freighters for whale meat in 1939, they produced
meat, fertilizer, leather and fibre from those parts of the animal
that were not a source of whale oil and which, in Antarctic Ocean
factory-ship whaling, were formerly discarded into the sea, except
for small amounts that were salted and stored for food (Ino 1940:
7–8; Baba 1942: 100–2, 106, 111–15). Even so, as the statistics in
Table 1.9 reveal, in this period the production of whale oil was
overwhelmingly dominant.[25]

Let us now consider the composition of the labour force involved
in these whale catching and processing operations. As I mentioned
before, at the time of the introduction of Norwegian-style whaling,
Norwegians always occupied the position of harpoon gunner. But
by this time, only one of the twenty-five whale catchers engaged
in coastal whaling had a Norwegian gunner. All the rest were
Japanese (Marukawa 1941: 126). When the Nihon Whaling company introduced its factory ship from Norway, however, it had
on board 'several Norwegians with experience in fishing ground
direction, whale catching, the processing of the whale carcass
on board the factory ship and the operation of oil extraction' in
factory-ship whaling in the Antarctic Ocean (Baba 1942: 98). No
Norwegian experts were employed on the *Tonan Maru*'s whaling
expedition the following year (Baba 1942: 98). But in the year after
that, 1936, when Taiyō Hogei took up factory-ship whaling in the
Antarctic Ocean, it employed a Norwegian gunner on its vessel, the
Nisshin Maru and again in its 1937 operations it had two or three
Norwegians in its crew (Kaiyō Gyogyō Kyōkai (Ocean Fishing
Association) 1939: 113). Further, when Kyokuyō Hogei joined
in factory-ship whaling in the Antarctic in 1938, it employed
six Norwegians (Iwasaki 1939: 56; Kyokuyō Hogei sanjūnenshi
henshū iinkai 1968: 144). On the other hand, Korean labourers had
been engaged in factory-ship whaling from the outset. According

to the *Mokpo Shinbo* of 24 April 1938, there were 30 natives of Ulsan and nine workers from Daehukdo island in Cholla Namdo assigned to work on the *Tonan Maru* (Park 1995: 324). Further, since it was possible to avoid being conscripted if you got a job with a whaling company, after this there were many who sought this kind of employment and by the end of World War II, there are said to have been as many as three hundred Koreans working in the whaling industry.[26]

Finally, let me summarise what we have clarified above regarding the process of introduction of factory-ship whaling. Firstly, we saw how it was necessary to combine the capital of large-scale enterprises with the personnel and facilities of existing whaling companies. Then, I made the point that the first step in the introduction of factory-ship whaling technology took the form of the purchase of a Norwegian whaling factory ship. But it did not stop there. Norwegians, who had pioneered factory-ship whaling in the Antarctic Ocean (See, for example, Baba 1942: 68–87) were also employed in the capacity of technical advisors. Further, it became clear that Korean workers had been engaged in factory-ship whaling from its inception, that at this time whaling had been positioned as a 'national policy' and that, unlike the situation at the time of the introduction of Norwegian-style whaling, the objective was now whale oil production.

Conclusion

Here I sum up my account of the introduction of technology into the modern Japanese whaling industry. The first point to note, in both the introduction of Norwegian-style whaling and with factory-ship whaling, is the fact that Japanese whaling companies employed Norwegians as the conduit for the introduction of the new technology. Next, we can say that there was a connection between the introduction of new whaling technology and pre-war expansionist policies. This is evident in the beginnings of Norwegian-style whaling on the Korean peninsula and in the 'national policy' of whale oil production at the time of the introduction of factory-ship whaling. And finally, as a result of the above two points, we can say that, in place of the social structure of the whaling group that was governed by heredity and status, a new social structure evolved based on national differences, in which Norwegians held high positions as bearers of the new technology, Japanese enjoyed privilege as managers and as inheritors of the

Norwegian technology and Koreans remained, for the most part, general seamen and process workers.

Given the process we have identified above, the only thing that we can really say was 'Japanese' about modern Japanese whaling was the fact that it was carried on by whaling companies managed by Japanese. Modern Japanese whaling was certainly not performed by *Japanese* alone. It was, in a variety of contexts, an industry that was largely run by people who were 'non-Japanese.' It was certainly not an industry that made use of technologies the Japanese alone possessed. For this reason, whaling as it was pursued by Japanese whaling companies cannot be categorised as 'Japanese.' That is to say, it was not an organic whole, but rather an amalgamation of various elements; nor was it a group-oriented phenomenon in contrast to 'the West.' In addition, we can identify a lack of continuity in the changes in social structure accompanying the introduction of the new technology, the shift from other industries into whaling management and, further, in the move to specialisation in whale oil production with the introduction of factory-ship whaling.

In short, the modern Japanese whaling industry developed against the contemporary background of expansionism, through an amalgamation of various people, divided according to their country of origin, employing a mixture of technologies, and as such it was completely different from the whaling that existed before. So, if we are to define 'culture' as having wholeness and continuity, when we consider the aspect of technological innovation, we can hardly claim that the case has been made for characterising Japanese whaling activity as 'whaling culture' or as Japanese 'traditional culture.'

2 Disturbance as an Interlacing of Experience: Analysis of the riot incident at the same station of the Tōyō Whaling Company

Introduction

The task of this chapter

As I mentioned in Chapter 1, after the introduction of Norwegian-style whaling, this method came to be practised, not only in the former net-whaling areas, but all over Japan even in areas which had not previously engaged in whaling, or, to put it another way, had not caught or made use of whales on the scale of the whaling groups. How then did the people living in those areas where whaling was newly introduced, particularly those engaged in the local fishing industry, feel when they found themselves confronted with whaling activity and how did they respond to it? This chapter attempts to explore these questions.

The reality is that we have widespread reports to the effect that people were not happy about the rise of this kind of whaling industry. In 1906 when Tōyō Fisheries started its operation in Chōshi in Chiba prefecture, the fishermen of the area gathered in a show of force and demanded the closing of the whaling station. Eventually, an ugly situation was avoided through mediation and whaling commenced under specified conditions, the details of which are not clear (Ōno 1907: 555–6). Further, when Tōyō Whaling established its station at Ushitsu in Ishikawa prefecture (year unknown), the fishermen of Toyama Bay opposed the development of a whaling industry there (Ayabe 1910; Matsuzaki 1910). Later, even in Ayukawa in Miyagi prefecture, one of the core whaling districts, a protest movement broke out when Tōyō Fisheries first penetrated the area in 1906. In the end, something of a settlement was reached through the good offices of the prefectural fisheries authorities, the head of Oshika county and the Ayukawa village administration under which Tōyō Fisheries agreed to donate three hundred yen a year to Ayukawa (Kondō 2001: 231–7).

The most violent expression of this kind of friction and confrontation, mainly involving local fishermen, occurred in Same village, Hachinohe County of Aomori prefecture in1911. The so-called 'Tōyō Whaling Company Same station riot incident' (hereafter abbreviated to 'the incident'), in which fishermen from Same village attacked and set alight the Tōyō Whaling station that had been established in the same year, was a major disturbance that resulted in injuries and fatalities.

In comparison to other disturbances this incident was so violent that, as we shall see from the following description, it has left behind a body of source materials that enables us to access the thoughts of those involved at the time. And, in so far as it may be seen as the peak of friction and confrontation over whaling, this incident gives us a clearer picture than any other of just how the local fishermen felt. Consequently, I would like to take this incident as a case study to throw light on the task I have set for this chapter.

Analytical perspective

There have been several previous studies of this incident. The most detailed research is the account of Ishida Yoshikazu (Ishida 1978). In addition, an overview of the incident can be found in local publications (Hachinohe Shakai Keizaishi Kenkyūkai ed. 1962: 199–219; Aomoriken Minseirōdōbu Rōseika ed. 1969: 103–13). Satō Ryōichi provides a guide to the available source materials (Satō 1987), and Kondō Isao includes the stories of people who were working in the whaling station at the time of the riot (Kondō 2001: 291–6).

It is true to say that this previous research, each with its own emphases, has given a summary of the incident, clarified the causes and provided background. That does not mean, however, that there are no problems over the way the incident has been presented. The problems to which I refer hinge on the fact that, in their desire to discover who was the ringleader and what he might have had in mind, previous researchers have paid little attention to the actual circumstances of the fishermen involved in the riot. That is to say, they have written up the incident on the premise that ringleaders, including some outside the fishing industry, had instigated the fisherman to riot, or on a belief that without some kind of instigation the fisherman would never have been galvanised into action.[1]

Here, rather, I would like to focus on the actual situation of the fisherman and to consider what it was that caused them to take collective action. In this regard, even if we assume for the sake of argument that someone from outside instigated the disturbance, we would still be left to consider how any such instigation would have been perceived by the fishermen.

In addition, I would like to analyse the incident in question, not as political or economic history as earlier researchers have done, but within the context of current sociological debates. Most sociological studies to date have dealt with disturbances from the perspective of social movements theory. And it has been generally accepted that there are three major elements relevant to the analysis of social movements; political opportunity structures,[2] resource mobilisation[3] and framing processes (McAdam, McCarthy *et al.* 1996; Takubo 1997). It is possible, indeed quite easy, to apply the first two of these three elements to the analysis of the incident in question. But to do so, it seems to me, would be simply analysing the incident once again in a similar context to previous studies. Consequently, in this chapter I have chosen to approach the subject from the viewpoint of those features that can be considered to fall within the definition of framing processes, i.e. 'the conscious strategic efforts by groups of people to fashion shared understandings of the world and of themselves that legitimate and motivate collective action' (McAdam, McCarthy *et al.* 1996: 6).[4]

However, in this chapter I shall focus on 'internal' rather than 'external' framing. 'External framing' here refers to the process in which a certain group, in order to gain supporters, argues its case to 'the world' outside the group. In contrast, 'internal framing' is when a group works to achieve a common understanding among those people who regard themselves as 'us.' Actually, when we divide framing in this way we come to understand that, largely due to the growth of the mass media and the difficulty of verifiable research, there tends to be a bias towards 'external' framing.[5] But if we are to consider the actual circumstances of the fishermen and what it was that made them group together, I believe that we need to examine their 'internal' framing.[6]

Where, then, should we turn our attention if we are to elucidate this internal framing? Here, we should note that the main focus of current research on social movements and disturbances, or the social movements debate since the resources mobilisation theory, has been on those studies that have identified the discipline and norms within social movements and disturbances, together with studies linking social movements and disturbances to everyday social ties. That is to say, interest has centred on the relationship between social movements or disturbances and everyday life.[7] To this end we can refer to studies in social history,[8] but we must be careful to consider the extent to which the actual examples we have chosen can be regarded as 'everyday,' given the site and timeframe of these people living their lives in communities engaged in fishing in early modern Japan.

Torigoe Hiroyuki, drawing on Japanese rural sociology and early modern history, used the term *seikatsu sekai* (life world) to describe the shared conceptual world of life consciousness based on each individual's life experiences. Further, he held that this life consciousness operated as knowledge that formed the basis for judgements when concrete action was required. This knowledge he called *nichijōteki na chisiki* (everyday knowledge). Further, Torigoe classifies this everyday knowledge into the three categories of; 1. individual knowledge of life experiences (not the experiences themselves but the way they are codified as knowledge), 2. common sense within a livelihood organisation (village, community etc.), and 3. popular morality introduced from outside the livelihood organisation (Torigoe 1997: 27–32). Of these, common sense is defined as 'the accumulation of wisdom of the livelihood organisation itself that makes the everyday lives of its members run more smoothly' (Torigoe 1997: 31). Torigoe goes on to explain popular morality as that body of concepts, created through the authority of the nation state, comprising diligence, thrift, filial piety, honesty and so on (Torigoe 1997: 29–31).

While we should bear in mind Torigoe's tripartite classification as we consider 'the everyday' in fishing communities in early modern Japan, in some respects his model requires further investigation. Is it true, for instance, that only popular morality is introduced from outside the livelihood organisations? Torigoe's reference to popular morality derives from an aspect of the research of Yasumaru Yoshio (Yasumaru 1965a), but Torigoe sees this as having been created by the State to control the people (Torigoe 1997: 26, 29–31). Whereas Yasumaru himself, in the body of his research overall (Yasumaru 1965a: ; 1965b), concludes that ultimately popular morality was a product of popular thought from the middle of the Tokugawa period (eighteenth century) developed to motivate efforts towards self-cultivation and self-discipline among the common people.

Further, he stresses the point that in the Meiji period this self-cultivation and self-discipline of the masses underpinned criticism of society, surfacing in the form of revolts and new religious movements. In this light we should probably consider that popular morality, whether consciously or unconsciously, not only supported the power of the State, but also at times functioned to oppose State power.

In addition, Torigoe's classification was fundamentally a model for static situations. Although he does point out in passing that common sense undergoes change (Torigoe 1997: 43), he does not clearly spell out how the dynamics of his three categories play out in the process of the penetration of popular morality, nor, indeed, touch upon how the categories relate to one another. In this chapter, however, we will need to mention these dynamic aspects, not least in the sense that we need to find out what changes occurred in the district as a result of the incident.

To summarise the argument to this point, our aim is to elucidate the process by which a group of fishermen, people who see themselves as 'us,' create a common understanding. We do so by focusing our attention on the actual situation of the fishermen themselves, the everyday knowledge they have acquired through their own individual life experiences and the way in which this everyday knowledge can change. Let us now proceed to look at the incident from this analytical perspective.

Case study

Overview of the Incident

First let me summarise the incident on the basis of previous research. 1909 seems to have been the year in which the Dai Nippon Whaling Company, founded in 1907, decided to establish a whaling station in the village of Same. The fisherman in general, objecting to the fact that they had not been consulted about the proposal that had been approved by directors of the fishing unions, stormed into the Sannohe County Office to voice their opposition to the plan. As a result of this action the various fishing unions withdrew their support for the establishment of the whaling station, and in the end the plan was put on hold when the prefectural authorities failed to approve the project. But some local men of influence, who advocated the establishment of the whaling station, had a whale carcass towed to the site and flensed there as a demonstration of the process. This action prompted a debate over whether or not the blood and oil discharged in butchering of the whale was harmful to the fishing industry.

Later, when Tōyō Whaling Company was created through the merger of Tōyō Fisheries, Dai Nippon Whaling Company and others (See Figure 1.1), a second plan was drawn up and an application made to the prefectural government to establish a whaling station. Despite a similar opposition movement against the proposal, the mayor of Same village privately petitioned the prefectural authorities, arguing that the establishment of a whaling station would promote the development of the village and finally, on 19 September 1910, the Minister of Agriculture and Commerce approved the building of the station.

The opposition continued, however, with village councils in the vicinity passing resolutions opposing the development and the director of the fishing union appealing directly to the Minister of the Interior. Later it was agreed the Same fishing union would be paid compensation of ten yen per whale, but no compensation was offered to other fishing unions in the vicinity.

Tōyō Whaling commenced operation at the station on 17 April 1911, planning to continue until the end of September. As its operations were running exceedingly favourably for the company, Tōyō Whaling applied to the Ministry of Agriculture and Commerce for a two-month extension, but without waiting for permission to be granted, i.e. in effect operating illegally, caught and flensed six whales in October of that year.

Early on the morning of 1 November, a group of fishermen, mainly from the district surrounding Same village, estimated variously at from several hundred up to one thousand, attacked the Same whaling station and set it alight. From there they went on to attack in turn the police substation, the inn normally used by employees of Tōyō Whaling, and the houses of those influential locals who had supported the building of the station. The riot finally subsided around noon. It is recorded that in the confrontation at the whaling station the casualties on the fishermen's side were; one dead, two seriously injured (one of these died later) and nine with minor injuries; four of the police were seriously injured, another four had minor injuries; three company employees were seriously injured and eleven more sustained minor injuries. (Ishida 1978: 285).

The roundup of the fisherman began the next day, 2 November. The determination of the preliminary hearing was announced on 6 December,[9] the public trial in the Aomori District Court opened on 5

February 1912 and sentences were handed down on 16 March. Seven of the defendants received prison sentences of eight years. A further nine were given six year sentences, seven two years sentences (of these three had additional suspended sentences of 3 years), seven received one year prison sentences (three of these had additional suspended sentences of 3 years) and six were fined forty yen each. Four defendants were acquitted and the trial of one defendant was delayed because of illness. Appeals were lodged against the sentences, but on the death of the Emperor Meiji all defendants were granted a general pardon on 27 September 1912.

The sources

My sources for the following discussion are mainly materials in the Hachinohe City Library available on microfilm under the title, *Kujiragaisha yakiuchi jiken kōhan kiroku* (Record of the public trial over the whaling company raid incident). It appears that the record was made by the office of Hanai Takuzō, a prominent barrister of the time, who had been engaged to defend some of the defendants in the case.

There are two rolls of microfilm, the first containing the transcripts of the interrogation of witnesses, descriptions of the crime scenes and medical reports on the dead and injured and the second containing copies of the depositions of defendants and the sentences. Of these source materials, I focus particularly on the statements of the defendants on the second roll of microfilm. As these materials record the testimony of those who actually took part in the riot, they can, I feel, enable us to approach the consciousness of the fisherman of the time. In the flowing discussion I refer to the first roll of microfilm as 'micro. A' and the second roll as 'micro. B.'[10]

In addition I used newspapers published at the time to get an idea of contemporary circumstances and the situation surrounding the trial. The papers consulted were the *Tōō Nippō*, the representative newspaper of Aomori Prefecture still published today, and the *Hachinohe* and the *Ōnan Shinpō* published in Hachinohe. But at the time of the incident these newspapers were the organs of political parties or political organisations. The *Tōō Nippō* was first published in 1888 as the official organ of the Daidōha (Unity faction) of the Liberal Party (Jiyūtō), which advocated a unified front for the party which was then heavily split into factions (Hachinohe Shakai Keizaishi Kenkyūkai 1962: 115). The *Hachinohe*, the first newspaper in Hachinohe was first published in 1900 as the *Hachinohe Shōhō*, changing its name in 1902. From around this time it became the official paper of the Doyōkai (Saturday Association) formed in 1889 from elements of the Jiyūtō and former samurai. Further, the *Ōnan Shinpō* was established in 1908 by the Ōnan faction of the Kōminkai (Citizen's Association) formed around the same time by merchants and supporters of the Kaishintō (Reform Party) to counter the influence of the Saturday Association (Hachinohe Shakai Keizaishi Kenkyūkai 1962: 110, 115–6, 147–8).

The *Hachinohe* – Saturday Association and the *Ōnan Shinpō* – Ōnan faction took up opposing positions on virtually every issue. The *Hachinohe* was in the camp of those supporting the establishment of the whaling station, while the *Ōnan Shinpō* was initially against the project, but later changed its opposition to support (Ishida 1978: 240–69; Hachinohe Shakai Keizaishi Kenkyūkai 1962: 204–6). Consequently, in using these newspapers as source materials we should be aware that each exhibits a degree of bias.

On the household finances of fishermen

First, before proceeding to the analysis of the case study proper, it is pertinent to explain here a little about the kind of lives fisherman of the day were leading. Accompanying the commodification of agricultural products from the last years of the Tokugawa Bakufu (mid-nineteenth century), there was a high demand throughout the country for sardine meal made from boiling up sardines. In this area (Hachinohe) the fisherman manufactured this fertilizer from sardines they caught in sweep seine nets. But because seine nets can only be used when schools of sardines come close to shore, the sardine fertilizer industry continued in the doldrums despite the high demand for the product. Then, at the end of the nineteenth century, an improved *aguriami* (a kind of round haul net) was introduced and a certain Hasegawa Tōjirō was able to achieve success by adding further improvements to the net. (Ishida 1978: 181–5, 224–8; Hachinohe Shakai Keizaishi Kenkyūkai 1962: 200–3).

As it happens, Hasegawa later was one of the local influential figures who was enthusiastic about the introduction of the whaling station (Ishida 1978: 235, 240–6, 253–4) and subsequently had his house damaged and his fishing nets and other property burned in the riot ('Same gyomin bōdō shōhō,' *Tōō Nippō*, 3 November 1911).

Let us drill down in a little more concrete detail on the lives of these fisherman who used seine nets and haul nets to catch sardines. Table 2.1. is a summary of the profiles of the defendants in the case (excluding those for whom no source materials remain) based on the depositions of defendants (including transcripts of the examination of witnesses, police interviews and the 'record of the conduct of offenders' (*hannin sokō chōsho*)), presumably prepared at the stage of the preliminary hearing and collected together in micro. B. First, the material gives us a clear picture of the fishermen's household finances at the time. For comparison I have prepared, in Table 2.2., an overview of the average income and expenditure for farming households in the same period. Further, we can compare these data with the results of a survey of the urban poor (*saimin*) conducted in Tokyo and Osaka in 1912 which reveals that the average monthly income for households struggling to survive under extremely harsh conditions was 28 yen 12 sen and expenditure was 28 yen 10 sen (Naimushō Chihōkyoku 1914: (1992: 612–19). Calculated on a yearly basis, this becomes an annual income of 337 yen 44 sen and annual expenditure of 337 yen 20 sen. We have no breakdown of the income and expenditure of fishermen and, while we might have reservations about grouping them with farmers and 'the poor,' a simple comparison of incomes shows that, on the one hand there were some individuals like L, whose annual income considerably outstripped that of a landlord farmer. On the other hand, from the data we can assume that several of the fisherman were leading far from affluent lives. In addition, given the disparities in annual income among the accused, i.e. those people whom the authorities judged to be the ringleaders of the incident, we can see that the group was made up of people drawn from a variety of social classes. It is also possible to confirm that there is no correlation between annual income and the severity of sentences.

The data in Table 2.1. thus confirms our view that the incident was instigated by the whole community.

Fishermen's accounts of the cause of the incident

Next I wish to discuss how the incident came about. Of course, this topic has already been taken up in earlier research, but rather than enumerating what has been said before, I should like as far as possible to clarify the position from the contemporary accounts of the fishermen concerned.

Table 2.1: The defendants

	A	B	C	D	E	F	G	H
Age at the time of the incident	33	39	44	49	51 (D.o.b unknown)	46	60	24
Occupation		Dried fish monger or fishing	Farming/ Fertilizer industry	Fishing	Fertilizer industry	Fishing	Sardine meal industry	Fishing industry
Village tax paid	3 yen 11 sen	36 yen 78 sen (Father pays)		34 yen 61 sen		8 yen 14 sen	12 yen 82 sen	
Annual income	None	Approx. 400 yen (Father?)	Approx. 1,500 yen	Approx. 500 yen			Approx. 200 yen	Approx 3,000 yen. (Household head)
Family	7 (Mother, wife, 4 children)	2 (Father)	4	10			5	
Military service	2nd reservist army private 1st class	None	None	Served in Sino-Japanese war			None	
Official appointments	Fishing union director	None	None	None		Fishing union councillor	None	
Sentence	Acquitted	2 years	2 years	1 year	6 years	6 years	6 years	8 years

	I	J	K	L	M	N	O	P
Age at the time of the incident	34	48 (D.o.b unknown)	39	64	45 (D.o.b unknown)	61	47 (D.o.b unknown)	29
Occupation	Fishing	Fishing	Fishing or unemployed	Fishing	Fertilizer industry	Fishing	Fishing	Fishing
Village tax paid	21 yen 28 sen (Household) head pays	43 yen 45 sen (Household)		294 yen 65 sen	24 yen 40 sen	16 yen 69 sen	15 yen 83 sen	48 yen 27 sen (Household head pays)

	I	J	K	L	M	N	O	P
Annual income	Approx.400 yen (Household head?)	Approx.1,500 yen (Household head)		10,000 yen	1,700 yen	500 yen	400 yen	Approx.1,200 yen (Household head?)
Family	7	7	1	9	10	3	7	11
Military service	Served in Russo-Japanese war, 2nd reservist army corporal (infantry)	None	None	None	None	None	None	None
Official	None	None	Former fishing union councillor	Village assembly councillor	Ward assembly councillor			
Sentence	8 years	6 years	8 years	2 years suspended sentence 3 years	2 years	2 years suspended sentence 3 years	2 years suspended sentence 3 years	Fine 40 yen

	Q	R	S	T	U	V	W	X
Age at the time of the incident	22	32	35 (D.o.b. unknown)	21	33	37	29	37 (D.o.b. unknown)
Occupation	Fishing	Fishing	Fishing	Fishing	Fishing	Fishing	Fishing Fertilizer industry	Fishing
Village tax paid	15 yen 70 sen				11 yen 50 sen (Household head pays)	(Unknown)	82 yen 53 sen (Household head pays)	
Annual income	Approx. 300 yen	Approx. 2,000 yen	None (In person), Approx.4–500 yen (Father)	Approx. 300 yen		(Unknown)		None
Family	4 (Household head plus three)	4	11		4	7	5	13
Military service	Served in Russo-Japanese war; army private 1st class (infantry)		None	None	Served in the Boxer rebellion (1900) Navy, able seaman (stoker)	None	Served in the Russo-Japanese war	None
Official appointments	None							
Sentence	Fined 40 yen	2 years	Fined 40 yen	Acquitted	1 year	Fined 40 yen	Fined 40 yen	Acquitted

Table 2.1: continued

	Y	Z	AA	AB	AC
Age at the time of the incident	31 (D.o.b. unknown)	46	41 (D.o.b. unknown)	26	27
Occupation	Fishing	Fishing	Fishing	Fishing	Fishing
Village tax paid	9 yen 60 sen (Household head pays)	1 yen 13 sen (Household head pays)			
Annual income	250 yen (Household head?)	(Unknown)	Approx. 4–500 yen	Approx. 500 yen	
Family	8	6	14	8	
Military service	None	None	None	Reservist, army, private (infantry)	Served in the Russo-Japanese war
Official appointments			Candidate for the village council		
Sentence	1 year suspended sentence 3 years	1 year	Fined 40 yen	1 year	Acquitted at the preliminary hearing

Notes:

1. Blanks indicate that there was no record in the source materials.

2. The order of the records follows that of *Kujiragaisha yakiuchi jiken kōhan kiroku*.

3. Of the defendants whose place of residence is recorded, all except AB are from the same administrative village.

4. Where the date of birth is recorded I have used that to calculate the age of the defendant. Where the sources record the age but not the date of birth I have given the age recorded and appended (D.o.b. unknown)

5. For family members I have followed the source materials, giving just the numbers when that is all that is recorded and adding details of family composition when available. I have used my discretion to calculate the number of family members including the defendant.

Sources: *Kujiragaisha yakiuchi jiken kōhan kiroku*, 1911a; *Ōnan Shinpō*, 10 December, 1911.

Table 2.2: Farming household income and expenditure (1911)

	Landlord	**Farming own land**	**Tenant farmer**
Household members	8	7	6
Income (yen)	4,097	804	613
Expenditure (yen)	3,435	738	584
Balance (yen)	662	66	29

Note: For landlords the average of 24 villages in taken, while for farmers working their own land and for tenant farmers the average is derived from data taken from 27 villages.

Source: Saitō, Mankichi (1918), *Nihon nōgyō no keizaiteki hensen*. Abridged version reprinted in Tada Kichizō ed. (1992), *Kakei chōsa shūsei 9 Meiji kakei chōsa shū*, Tokyo: Seishisha, pp. 224–54.

In the 'record of the conduct of offenders,' presumably compiled by the police authorities, under the heading 'motivation for the crime' there are, for example, a number of accounts like the following.

He is one who participated at the scene after being instigated to riot by (name omitted) who brought up the idea of the superstition that whale oil and blood are detrimental to fishing as the motivation for the action. (From the statement of defendant B in Table. 2.1. as recorded in Micro. B.)

This account, with a few minor differences in expression, is common to almost all the statements of the accused noted in the 'record of the conduct of offenders.' That is to say, it is the 'official' reason put forward by the police for the fishermen's riot.

In contrast to this, what did the fishermen have to say about the causes of the incident? To help answer this question I have set out below, wherever I felt it had some bearing on the motive for the riot, all the responses of defendants to the magistrate's question, 'are you opposed to whaling?'

1. Defendant F (part 1)
   Question: Is the witness in favour of or opposed to whaling?
      Reply: I realise we cannot tell whether it is harmful or not without an investigation, but nevertheless somehow I feel opposed to it.

2. Defendant F (part 2)
   Question: Did the defendant on 31 October meet with others at Tatehana and discuss the attack on the whaling company?

Reply: I go out to Tatehana every morning and evening to check on the weather, but I certainly did not discuss anything like attacking the company and I have not been present anywhere when that kind of thing was being discussed.

I did say that I'm against whaling, because even if we don't know whether it is harmful or not, there are no fish about and that could be on account of the whaling.

3. Defendant G

Question: Is the witness against whaling?

Reply: I am neither for it nor against it.

4. Defendant I

Question: Does the witness support or oppose whaling?

Reply: I am against both whaling and flensing. The sardines don't come in if there is whaling going on (the exact meaning here is unclear as one of the characters in the original transcript is illegible) and the blood and oil from the butchering of the carcass kills the fish.

5. Defendant L

Question: Is the accused against whale flensing?

Reply: I am one of those who put in a petition to have it stopped because we don't get any catches when there is blood and oil discharged into the sea.

6. Defendant W

Question: Is the witness against whaling?

Reply: Whales are gods. It's bad to catch them.

The mention of Tatehana in F's second response above is a reference to the 'ocean rescue station' (*suinan kyūsaijo*) situated in a place of that name. It is not clear what was discussed when fishermen met here on 31 October or how the decision was reached. But it was taken for granted that the decision to move on the Same whaling station was taken here at a meeting of the fisherman and the magistrate at the preliminary hearing asked most of the defendants if they had participated in the meeting (Micro. B.).

Returning to the cause of the incident, the sources show that only five of the defendants mentioned the reason for the action. Even so, we notice a range of responses from vague to very clear opposition to whaling. Of course, those being questioned were under investigation and may have consciously adopted an equivocal attitude (according to the transcripts of the depositions of defendants, all the accused denied participating in the riot) and we can consider that among the fishermen there was a variety of reasons for taking part.

Nevertheless, we need to note that, as recounted above, clearly in the minds of the fishermen, opposition stemmed from the fact that the blood and oil from the whale flensing process either killed the fish or stopped them from coming close to shore or from a belief that whales are gods and it is wrong to kill them. Let us now proceed to analyse these two points in more detail.

Ebisu and pollution – analysis of the accounts

First, the point that whales are gods and must not be killed is by no means merely a figment of the imagination of defendant W in Table 2.1. This is clear from mention of this topic in the written judgement of the preliminary hearing. The original Japanese is rather long and rather difficult to read as it has no punctuation marks, but this should not be a problem in the English translation quoted in full below.

> Originally, in the district of the defendants, whales were called o-Ebisu-sama (Revered Mr Ebisu) and regarded as sacred. It was held that sardine fishing depended a great deal on the benevolence of passing whales and there was a custom in the area whereby, as soon as a whale spout was seen far out to sea, those watching would clap and bow three times in prayer beseeching the god for good fortune in fishing. Consequently, there are many traditional tales and proverbs about how shoals of sardines coming close to shore are blessings from the god Ebisu to the fishermen living along the coasts. And, because there are still some among the fishermen even today who believe this, any talk of catching whales, let alone cutting them up and letting their blood and oil spill into the sea, is regarded as anathema to them. Further, they have a fear that whaling ships will prevent fish from coming in or scatter the shoals. (following omitted) ('Same bōdō yoshin shūketsu,' *Ōnan Shinpō*, 10 December 1911a)

This indicates that the magistrate in the preliminary hearing recognised the fact that in this area it had long been the custom to worship whales as gods that brought blessings in the form of sardine catches and that this was why they were against whaling. It seems to suggest that, although the fisherman other than Defendant W have left no mention of this fact in the written record, there must have been similar responses in the exchanges between the defendants and the magistrate.[11] Further, it is reported that the belief that whales are incarnations of the god Ebisu stems from the fact that sardines, the principal catch in this area, are driven into shore by whales pursuing them for food ('Hogei bōdō jiken shokan,' *Tōō Nippō*, 12 December 1911).

What are we to make of this kind of reasoning by the fishermen? If we follow Torigoe's classification of everyday knowledge, we can think of the fishermen's belief that whales are gods that bring blessings in sardine fishing as having been formed in the repeated daily activity of fishermen catching sardines, to the point where, *at this particular point in time*, it had become a matter of common sense. It will soon become clear that common sense as everyday knowledge acts to create a common understanding in which it is the fishermen themselves who decide about whether or not to participate in the riot.[12]

Let us now consider the other reason, namely that the blood and oil released in the flensing of the whale carcass kills or drives away fish. It is certainly possible to see this as an extension of the fact that whales were worshipped as gods. And there is also a hint of support for this interpretation in the police authorities' dismissal of this claim as a 'superstition.'[13] But the debate over whether or not blood and oil are harmful had been going on in this area from before the time of the riot (Ishida 1978: 240–54). In addition, it should be noted that no clear ruling is made, in either the written judgement of the preliminary hearing (1911a, *Ōnan Shinpō*, 10 December and 1911b, *Ōnan Shinpō* 13 December) or in the sentences of the trial[14] on the question of whether or not whale blood and oil are harmful.

Now let us look at the evidence from the record of the interrogation of the head of the Same whaling station and the managing director of Tōyō Whaling, included in Micro. A., to clarify whether the blood and oil discharged in the flensing of the whale carcass is indeed detrimental to fishing. In evidence they explained that the flensing process at the Same station was carried out with the whale carcass drawn up completely onto dry land (See Chapter 1) and that the blood was temporarily retained then pumped off to a holding pond. However, they acknowledged that during this same period, the company had been cautioned by the prefectural authorities regarding its handling of blood and oil. And with considerably more whales caught than had been predicted for Same – 37 taken in one week in June alone – they had to admit that their processing facilities were inadequate and that the company, unable to cope with the volume, had discharged blood into the sea. In fact, from the beginning of April to the end of October 1911 a total of 263 whales was caught by the Same station. This number was made up of 240 fin whales and 23 sei whales (3 fin whales and 3 sei whales were caught in October). This is said to be a record for the number of baleen whales taken by a single whaling station in one year (Kondō 2001: 292–3). Moreover, given that we have a description of approximately 300 whales being flensed at the Same station (Hachinohe Shakai Keizaishi Kenkyūkai 1962: 206), it is quite possible that other species may also have been caught. According to Kondō Isao, due to the large volume of whales caught, the fertilizer dealer who had the rights to buy waste from Tōyō Whaling, was unable to cope, so this was acquired by connections of the local men of influence who had supported the building of the station (thus excluding any who opposed the station from sharing in the profits) and manufactured into fertilizer. In addition, the large number of whales caught meant that great quantities of blood and whale oil, in addition to other waste liquids from the processing operation ended up being discharged into the sea. Kondō, who was working in a coastal whaling company, reports that, when a large volume of baleen whales had been butchered and their carcasses discarded without further processing, the blood would coagulate and settle like mud on the sea bottom to a depth of about three metres and that when large quantities of the blood and waste from the boiling process were discarded, depending on the profile of the seabed, it sometimes caused the death of marine life through lack of oxygen and it completely annihilated surf clams (*hokkigai* (*Spisula sachalinensis*))(Kondō 2001: 292–4).

From these reports it seems reasonable to assume that there was deterioration in water quality as a result of butchering of whale carcasses. In our modern terminology the Same whaling station was a source of pollution and a public nuisance. Consequently, the fishermen's conviction that fish were being driven away or killed by the blood and oil from the butchering of whales had emerged from the conditions that they themselves were actually experiencing. And we can see how that fact, together with the fisherman's belief that whales were gods that blessed them with catches of sardines, just happened to fit perfectly with the premise 'we can no longer catch sardines because of the water pollution resulting from the taking of whales which are divine creatures' and came to assume the form of a shared consciousness to oppose whaling.

The significance of military experience

There is one more factor we need to consider in our investigation of the fishermen's way of thinking. That is the fact that the list of defendants in Table. 2.1 includes a fixed proportion of men with military experience. As is also apparent from newspaper articles on the public trial ('Same bōdō jiken kōhan,' *Tōō Nippō*, 6 February 1912; 'Same bōdō jiken kōhan' 1912),[15] at least eight of the 41 defendants had experienced military service. And it was reported the fisherman who died immediately at the scene had served in the army during the Russo-Japanese War (*Tōō Nippō*, 3 November 1911; *Tōō Nippō*, 11 November 1911). ('Same gyomin bōdō shōhō,' 1911; 'Gyomin bōdō yobun,' 1911).

In the arson attack it is claimed that first an 'advance party' of seven men (including the two fishermen who were killed) headed for the station. Of these, at least two had had military experience (Ishida 1978: 284; Satō 1987: 25). In addition a 'fisherman's death squad' (*gyomin kesshitai*) of from twenty to thirty men was formed. It was reported in the press that this death squad clashed violently with the police and commanded the other fishermen (*Hachinohe*, 7 November 1911; *Tōō Nippō*, 3 November 1911).

From these facts we can tell that the riot was deliberately planned and carefully organised. In this regard, some have expressed the opinion that perhaps even at this time, experience of the so-called *hyakushō ikki* (peasant rebellions) had been handed down among the fishermen (Hachinohe Shakai Keizaishi Kenkyūkai 1962: 209; Ishida 1978: 275–82). It may be necessary to note this claim from the viewpoint of the nexus between 'tradition' and disturbance (see Note 7 of this chapter), but this cannot be confirmed from the sources at hand. We cannot rule out that possibility, but it seems more natural to assume that those with military training and war experience, used that experience to organise and command the fishermen and themselves lead the combat activity.

As the fishermen were not professional soldiers they should have had military training as conscripts, the Conscription Ordinance having been issued in 1873. Becoming a soldier even as a conscript is not simply a matter of learning how to use a gun and how to fight. It entailed complete physical training far beyond 'how to fight,' not only on the battlefield, but in the way you walk or behave, in your bearing and manners, your every movement. In a sense the individual was required to change. On the level of consciousness it required, without going so far as to quote here from the Imperial

Rescript to Soldiers and Sailors (1882), a will to give your all for the sake of the State ('*kokka no tame*'). This meant that the morality to support the power of the State permeated the populace through the instrument of military service. You might say that this was one of the circuits through which popular morality, comprising an array of concepts like diligence, thrift, filial piety and honesty came to take on the form of everyday knowledge for the common people.

Nevertheless, the modern whaling industry in Japan, from the time it began to receive financial support under the Law for the Promotion of Pelagic Fisheries, promulgated in 1897 (Ishida 1978: 42–6), and, as I argue in this book, not only at the time of its introduction (Tōyō Hogei K.K. 1910: 195–6), but throughout its history, has always been under the protection of the State. And in the area around the Same whaling station it was asserted that the whaling industry was not only to the benefit of that district but to the State as a whole (See, for example, 'Hogei hantoshi no kansatsu' (Half a year's observation of whaling), *Hachinohe*, 1 October 1911). And to raise a riot is to challenge the authority of the police and disturb society at large. In short, the arson attack on the whaling company was partly a matter of an objection to the State and its activities being expressed in action.

When we consider the planned and organised nature of the attack, it seems unlikely that the fishermen, particularly those among them who had had military experience, would have participated so positively in the incident and taken the lead in carrying out the attack if they had seen themselves as 'subjects of the State' (*kokumin*), that is to say if they had formed their subjectivity around discipline; if they had been individuals who themselves autonomously participated in their control by the State (Fujitani 1994: 170). If that is indeed the case, why, we may ask, was this so?

If those returning to their communities from military service simply went on living their normal daily lives, the experience they had gained as soldiers would not extend beyond their own individual circumstances. So we can imagine that, in order to ensure that those with military experience became *kokumin* and went on being *kokumin* some kind of medium was required, that is to say a group where they could mutually confirm and have confirmed their status as *kokumin*. In fact, we can see that men with military experience associated with one another and that there was a returned soldiers' group, the so-called *zaigō gunjin* (literally, 'soldiers in their home towns') in the area (in the 'Record of the conduct of offenders' for

Q and AB in Micro. B.). But the function of that network in the incident remains unclear.

Next, considering the general run of fisherman listed as defendants in Table. 2.1, it is clear from the sources that some of them were unable to write (Micro. B).[16]

As it was not long after school education, one of the most important instruments of nationalisation policy (*kokuminka*), had been introduced (The Education Ordinance had been issued in 1886 and the Imperial Rescript on Education in 1890), we can surmise there were some people in this district who had had no opportunity to receive an education or had been little influenced by one.

But rather than these speculations, and leaving aside the question of whether or not the popular morality motivated efforts towards self-cultivation and self-discipline, we can perhaps at least understand how the concepts of diligence, thrift, filial piety and honesty in the minds of the fishermen in general and those with military service in particular, underpinned the movement in which they literally put their own lives on the line to confront the power of the State for the sake of their community and comrades. In short, nationalisation policy was not functioning. It was the common sense *of that particular point in time*, and the norms associated with it, that held sway. Only 'the way to fight' was adopted out of the context of nationalisation. Further, it is possible that the popular morality that had existed from before conscription and school education, combined with nationalism infused through military service as the morality underpinning state power, had paradoxically functioned to support confrontation against the State.

Harada Keiichi, in the context of his investigation of memorial services for the war dead, points out that the emphasis on spiritualism and 'loyalty to the Emperor and love of country' (*chūkun aikoku*) comes after the Russo-Japanese War. He goes on to relate this fact to the way monuments to the war dead *chūkonhi* (literally, 'loyal souls monuments') appear after the Russo-Japanese War and increase dramatically in number after World War I. In other words, the various memorials all over the country that celebrated those who had fulfilled their duty to the State, including those who had returned alive, were subsumed under the category of *chūkonhi* (Harada 2001: 243–50). Harada states that 'from the time when only *chūkonhi* came to be built, the gaze of the people was compelled to focus on another emperor beyond the State' (Harada 2001: 243–50). From this time, to create subjects of the nation it was necessary to imagine the emperor as a divine presence extending beyond the State.

While not disagreeing with this, given that the morality supporting state power instilled through military service had not penetrated that far in the case of the war veterans who participated in the riot, I also admit the possibility of a slippage in ideology.

The aftermath of the incident

As I mentioned above, a degree of closure was brought to the incident when all those found guilty were pardoned on the death of the Emperor Meiji. But what were the subsequent developments of whaling in the area and how did the locals respond to them? Finally, I would like to try to shed light on these questions, which have received very little attention in previous research.

On 16 October 1912, about three weeks after they were pardoned, those who had gone on trial and some of their associates held a memorial service for the Emperor Meiji at a local temple. In addition to the former defendants, the prosecutor of the Aomori District Court, the prosecutor of the Hachinohe Ward Court, the head of the Hachinohe Police, local men of influence and others, between one hundred and fifty and two hundred people in all, attended the service. There, first priests of the various sects in the district read sutras and the participants paid their respects by burning incense. There followed a 'spiritual talk' (*seishin kōwakai*), in which the mayor of the rioters' village, after his opening remarks, went on to read a formal printed admonition (*kun'yu*) from the chief public prosecutor at the time of the granting of the pardon, and exhorted the men 'to make it your motto recalled morn and night' (*Ōnan Shinpō*, 19 October 1912). In addition, the prosecutor of the Aomori District Court spoke on 'Show the truth of your gratitude,' the prosecutor of the Hachinohe Ward Court on 'The terrible problem of crime' and the head of the police in Hachinohe on 'Practical ways to show gratitude.' Then, apparently, a representative of the temple organisation went on to discuss the topic of loyalty and patriotism (*chūkun aikoku*) and a representative of the local men of influence spoke on 'The spirit of unity' ('Gyomin Shaonkai,' *Ōnan Shinpō*, 19 October 1912; 'Jūōin no Hōtōkai,' *Hachinohe*, 19 October 1912).

The most detailed account we have of these speeches is that of the prosecutor of the Aomori district court ('Shaon no jitsu o ageyo (ichi),' *Hachinohe*, 19 October 1912; 'Shaon no jitsu o ageyo (ni),' *Hachinohe*, 22 October 1912; 'Takagi kenji no kunwa (ichi),'

Ōnan Shinpō, 19 October 1912; 'Takagi kenji no kunwa (ni),' *Ōnan Shinpō*, 22 October 1912; 'Takagi kenji no kunwa (san),' *Ōnan Shinpō*, 25 October 1912).

In his speech, as reported in the press, he says that, while one should never forget the sense of gratitude, it is also necessary to think of practical ways in which one can repay a debt of gratitude and spells out in point form a number of suggestions the defendants should bear in mind. In order to respond to 'Imperial benevolence' (*seion*), he lists a number of duties one must fulfill, including, obeying the law of the land, performing military duty and paying taxes. Then he goes on to say 'by constantly keeping in mind these points I have mentioned and expressing them in practical action, each of you can become loyal subjects and thus repay the vast divine benevolence bestowed upon you' ('Takagi kenji no kunwa (ni), 1912).

In all this we can see the strong desire of those in positions of power to use the amnesty granted on the death of the Emperor Meiji as a means to accomplish the task of turning the fishermen into subjects of the nation. Of course, it is unlikely that the fisherman took everything they heard at this gathering at face value and assiduously practised the precepts each day. But at the same time we cannot deny that perhaps there were some among them, chastened by the experience of arrest and imprisonment, for whom these messages struck a chord deep in their hearts. Further, we need to be aware that, in the context of the incident to this point, being loyal subjects, had become inextricably linked to an acceptance of whaling.

With this in mind, let us now move on to consider how whaling developed. Tōyō Whaling applied to Aomori prefecture on 31 May 1912 to reopen its whaling station, inaugurated its rebuilt plant in mid-June and put its whaling ships to sea on 29 June. Apparently, over this period the manager of Tōyō Whaling, Oka Jūrō, was engaged in frequent discussions with local stakeholders (Ishida 1978: 326). On 10 July 1912, Tōyō Whaling signed a written memorandum of understanding with members of the fishing unions in the area specifying the following points: 1. The company admits in writing and in testimony to the court that, although the station was built to handle one hundred whales it processed approximately three times that number, with the result that blood from the flensing process was discharged into the sea. Further, the company acknowledges that its continued whaling after the expiry of the period of its licence caused distress to the fishing community; 2. That the company will donate

Figure 2.1: Number of whales processed in Same

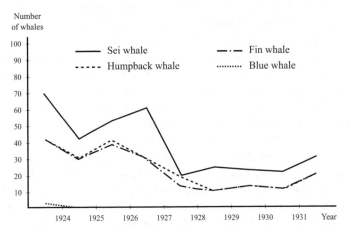

Notes

1. There is insufficient data to complete the graph for some species where the numbers taken were very small; often no more than one or two whales per season.

2. Other species, including the sperm whale, grey whale and right whale were taken sporadically and in such small numbers that they do not appear in the graph.

3. As only fragmentary data is available prior to 1922 it is not included in this chart.

Source: Kondō, Isao (2001), *Nihon engan hogei no kōbō, San'yōsha*, 293; Nōrin Daijin Kanbō Tōkeika, 1926–33, tables 1–9; Nōshōmu Daijin Kanbō Tōkeika (1925), 'Dai yonjūji nōshōmu tōkeihyō'.

funds to meet the cost of the trial; 3. While the offenders are serving their sentences the company will, as far as possible, employ members of their families in the whaling station; 4. While unfortunately it must exclude those already established in the industry, the company will help facilitate the setting up of other enterprises related to whaling; 5. While waiting for 'whaling industry workers' to be trained in the trade, the company will gradually, employ workers from this area (Ishida 1978: 326–7).[17] Further, Oka is said to have apologised to those involved in the incident for the inadequacy of the company's facilities and for failing to comply with the time limit on his whaling licence (Aomoriken Minseirōdōbu Rōseika (Division of Labour Administration 1969: 111); Satō 1987: 37). In assessing this response, Ishida Yoshikazu stresses Oka's virtuous character, but, given the extremely large number of whales they had caught, a more convincing reason might be, as Ishida points out in the same passage, that the prospects for whaling in the vicinity of the Same station were very promising indeed (Ishida 1978: 326–8).

So we have seen how whaling was resumed in this area, but twenty years later, this time at the company's instigation, the Same whaling station closed down. A newspaper reporting at the time on the reasons for the closing of the station, wrote as follows.

They say the reason for the closure is the fact that, though there are whales to be taken, there is no profit in the industry. They cannot make a profit because the prices of whale meat, whale oil and whalebone meal have plummeted in the depressed economy, but there is another hidden reason (Tōyō whaling stops work at its Same station, *Ōnan Shinpō*, 13 March 1932).

Actually, in the 1930–31 season, with the development of whaling in the Antarctic Ocean, the worldwide production of whale oil reached a record 614,496 tons. But under the impact of this over-supply, on top of the effects of the worldwide depression, the price of whale oil fell sharply, to the point where only five whaling factory ships set sail for the Antarctic Ocean in the 1931–32 season compared with forty-one ships the preceding year (Baba 1942: 78–84). And Tōyō Whaling itself in 1930 decided for the first time in the company's history not to pay a dividend (See Chapter 1). In addition, one could also point to the fact, clearly shown in Figure 2.1, that they were now catching far fewer whales in this area than they had in the early stages of the operation.[18]

In addition, contemporary reports in the local press indicated that around five hundred people were put out of work with the closure of the Same whaling station and that there was also a major impact on businesses in the area that relied on the custom of workers at the station. One paper reported that 'in marked contrast to years gone by when people objected to the building of the station on the grounds that it would harm fish stocks, the local people now hope that a way can be found for the whaling station to continue to operate for the sake of local employment' ('Tōyō whaling stops work at its Same station,' *Ōnan Shinpō*, 13 March 1932). Beyond this we have no way of knowing what the feelings of the fishermen might have been. We can fully imagine, however, how the livelihood organisations of the fishermen of the area and their common sense had changed over the intervening twenty years in which they were employed by the whaling station and the catching and flensing of whales had became an everyday occurrence for them.

The available statistics tell us that from this time up to the beginning of World War II, only one whale, a sei whale taken in 1933, was caught at Same (Nōrin Daijin Kanbō Tōkeika 1926–30, 1932–36, 1938–42).

After the war, in the period from 1947 to 1949, we can see from the statistics that a small number of whales were taken at Same (Nōrinshō Nōgyō Kairyōkyoku Tōkei Chōsabu 1949–51),[19] but the station closed down again in 1949 (Maeda and Teraoka 1952: 111).[20]

Summary

We have taken the incident of the riot at the Tōyō Whaling Company's Same station as an example of how fishing communities felt and how they reacted when faced with the prospect of taking whales. Before closing this chapter let me summarise what has emerged from the analysis of this incident.

It is obvious that the station was a source of pollution and was doing harm to fishing communities in the area. This must have been a particularly heavy impost on those several fishermen who were, we can assume, living close to the poverty line. On the other hand, in this place at this particular time it was common sense, knowledge formed in the repeated everyday routine of people catching sardines together, to see whales as gods that brought blessings in sardine fishing. These two facts combined to form the logic that 'as a result of catching whales, which are gods, the water becomes dirty and we can no longer catch sardines.' This notion came to be shared among the people as a conscious opposition to whaling. Then, under the added impact of a number of events and, finally, the discussions on the day before the riot, the situation tipped over the edge.

In addition, in relation to the consciousness of the fishermen, we noted the point that several of those who came to trial had had military experience. However, since the attack on the whaling station seems to have been carried out in a planned and organised manner, we can conclude that these men with military experience might be described as soldiers, but they had not reached the point of becoming subjects (*kokumin*). This suggests the possibility that the popular morality, and the morality to support the power of the State, instilled through military service, can actually function to promote confrontation against the State.

But the riot did not put an end to whaling in this area. We can assume that, as the catching and butchering of whales became part of the everyday routine, major changes occurred in the fishermen's notion of common sense. We saw too, how those in authority sought to use the riot and the amnesty granted on the death of the Emperor Meiji to turn the fishermen into loyal subjects (*kokumin*).

We concluded, in the context of the incident in question, to become a *kokumin* implied an unconditional acceptance of whaling. In short, in this area until the whaling station appeared on the scene, popular morality functioned alongside the common sense that whales were gods that blessed the people with catches of sardines. I think we can say that as a consequence of the riot, this way of thinking was replaced by the morality that put the Imperial State above all else, in combination with a new common sense formed from the catching and butchering of whales becoming a matter of the 'everyday.'

At any rate, in this chapter we have seen how the interlacing of the experience of the fishermen led to a riot. But, in so much as some aspects of experience can be constructed, we have also seen how the mix of experience stopped subsequent riots from occurring.

3 The Designation of Whale Species as Natural Monuments: How far does industrial use come into consideration?

Introduction

The purpose of this chapter is to elucidate the history of the protection system as it pertains to whales. In doing so, I consider the history of protection of whales in the context of the development of the system of natural protection itself, particularly that concerning the designation of natural monuments (*tennen kinenbutsu*). And finally, I would like to add some thoughts of my own regarding the current system, and various other strategies now being tested, for the protection of wildlife. That is to say, I think, by taking a historical approach, I can perhaps contribute a few effective arguments about the present system of wildlife protection in general, to add to the debate over protection of whales which has become more prevalent and more heated through its association with the so-called 'whaling issue.'

With this in mind, let me first confirm the categories of animals protected under their designation as natural monuments and then go on to consider the process and detail of the designation of whale species – for practical purposes just the finless porpoise and the grey whale – as natural monuments. Then, I shall append my own opinions formed in the light of the historical investigation.

Natural Monuments Conservation Schedule

Now, before launching into the analysis of concrete examples, let me begin by discussing the various connotations applied to the idea of protection (*hogo*), when used in the sense of the protection of wildlife, as we have been doing since the beginning of this chapter. Morioka Masahiro points out that there are two schools of thought

Table 3.1: Schedule of categories for the conservation of historical sites, scenic places and natural monuments

The following is a summary of categories approved for conservation as historical sites, scenic places or natural monuments. (The examples given in the present schedule have been chosen for convenience of explanation. This does not necessarily imply that they will be designated.)

(Section omitted)

Natural Monuments
The following have been recognised as warranting conservation as natural monuments

Section I
1. In relation to animals, the following categories have been approved for conservation

    i. Those significant species that are currently found in Japan and have not yet been discovered elsewhere in the world (e.g. the mikado pheasant (*Syrmaticus mikado*) and the fire-breasted flower pecker (*Dicaeum ignipectus*) of Taiwan, the Ryūkyū robin (*Erithacus komadori*) of the Ryūkyū Archipelago, and the Amami jay (*Garrulus lidthi*) and native black rabbit (*Pentalagus furnessi*) of Amami Ōshima island, etc.).

    ii. Those species that until comparatively recently lived in other parts of the world, but have subsequently decreased in number to the point where the small populations that remain can only be found in Japan (e.g. the grey whale (*Eschrichtius robustus*) of the Japan Sea).

    iii. Those species that live in Japan or its territorial seas that have in recent years been on the brink of extinction (e.g. the spoonbill (*Platalea leucorodia*), the great egret (*Ardea alba*), the Japanese crested ibis (*Nipponia Nippon*) and the great bustard (*Otis tarda*), the sable (*Martes zibellina*) of Hokkaidō and Karafuto (southern Sakhalin), the musk deer (*Moschus moschiferus*) of Karafuto, the Kurile sea-otter (*Enhydra lutris*), the Karafuto fur seal (*Callorhinus ursinus*), the Okinawan dugong (*Dugong dugon*) etc.).

    iv. Those species, the conservation of which is desirable because, though not unique to Japan, they are of considerable significance in East Asia (e.g. the giant salamander (*Andrias japonicus*) etc.).

    v. Breeding grounds or migratory sites of significant species (e.g. migratory sites of cranes in Yashiro village of Kumage county, Yamaguchi prefecture and Akune of Izumi county, Kagoshima prefecture and the oriental stork (*Ciconia boyciana*) breeding grounds of Tsuruyama, Izushi of Hyōgo prefecture, the breeding grounds of the black-tailed gull (*Larus crassirostris*) on Kabushima, Hachinohe, Aomori prefecture, the breeding grounds of the streaked shearwater (*Calonectris leucomelas*) on Birōjima of Kōchi prefecture, the sea along the Etchū coast of Toyama prefecture, site of the massing of luminescent dwarf squid (*Watasenia scintillans*), the breeding grounds and migration sites of birds etc.)

    vi. Fossil remains of various giant mammals, such as the elephant, rhinoceros and deer, and other significant animals found throughout Japan and the sites where they were discovered.

Table 3.1: continued

vii. Animals and groups of animal species as a whole unique to habitats such as, mountain, plain, marshland, forest, wetland and lake, seashore, river and sea, island, cave etc.

viii. Domesticated animals unique to Japan (e.g. the Tosa long-tailed fowl, the Japanese rumpless bantam, the Japanese bantam, the Japanese chin, the Tosa dog, the Akita dog, the Oki horse, the Tosa pony, the 'cowhorse' (so-called because it had virtually no mane or tail) of Tanegashima etc.)

ix. Significant species other than domestic animals introduced into Japan from other countries that are now in a wild state (e.g. the common pheasant (*Phasianus colchicus*) of Tsushima Island, the Hizen magpie (*Pica pica*), and the Ogasawara (Bonin Islands) deer (*Cervus unicolor boninensis*))

(Following omitted)

Source: *Kanpō*, No. 2258, 16 February 1920.

in our logical responses to the question, 'why should we protect nature?' These are *conservation* (*hozen*), which conveys the sense of 'protecting nature for the sake of human beings' and *preservation* (*hozon*), which means 'protecting nature for nature's sake.' Taking these contrasting axes of conservation and preservation, Morioka then adds the opposition of whether one is for or against human intervention in the development and regulation of natural protection, to arrive at a fourfold analytical frame (Morioka 1999: 32–48). Here, however, I would like to stress the fact that we can divide the idea of protection of nature into the concepts of *conservation* and *preservation*. Within this distinction, as will soon become clear, this chapter probes the logic of the conservationist ideology in natural protection. This approach sees wildlife as a 'resource,' and strives for continued exploitation of a particular natural resource on a relatively large economic scale. It follows, then, that this 'resource' needs to be protected. It is this utilisation of wildlife species as an industry that I particularly want to bring under the microscope in the following analysis.

What, then, was the logic behind the establishment of a system of natural monuments in relation to the protection of wildlife in general? Table 3.1 is the section on animals, extracted from the *Official Gazette* (*Kanpō*), No. 2258, published on 16 February 1920.[1] The list itself, decided on 28 December 1919, contains all items to be protected under the Law for the Conservation of Historical Sites, Scenic Places and Natural Monuments that had been passed earlier the same year. From this list we can see that certain animals indicated as 'significant' or 'unique' and others with declining numbers are designated natural monuments.

It is very difficult to ascertain the reasons for and aims of wildlife protection that must underlie the designation of natural monuments. In this regard, let us now turn to the works of the zoologist Watase Shōzaburō (1862-1929) (Miyoshi 1929: 376–80), who was deeply involved with the designation of natural monuments through his participation in the debate over draft legislation and the drawing up of the list of species to be protected.[2] In his oral account, (Watase 1921a; 1921b; 1921c; 1921d), Dr Watase summarises his aims for 'restoring the natural world,' through the establishment of national parks and wildlife protection refuges, under the following three headings; '1. the conservation of academic research materials, 2. the need to protect and propagate the resources of the natural world so that they do not become extinct, and 3. to protect areas where we can enjoy the beauties of nature and pass these on to later generations so that they too may benefit from nature's blessings' (Watase 1921d). And, particularly in regard to the third point, he says, 'this stems from the fact that from around the end of the nineteenth century and into the twentieth century, we have seen, on the one hand, a deal of self-examination and, on the other, opposition to the unfortunate trend towards material civilisation with its destructive technology' (Watase 1921a). Further, he notes 'through the application of science … we shall return nature, now on the brink of devastation, to its former state, or even produce better results than were achieved in the past' (Watase 1921c). As an example he cites the fox farming industry,[3] in which large profits were generated through the farming of wild foxes. Further, he goes on to state,

> It seems likely that with a thorough research effort we may well find species among the animals living today that will be beneficial to us in the future… It would be most unfortunate for us if any of these wildlife candidates were to die out before we were able to investigate them (Watase 1921d).

Despite his reflection on material civilisation spawned by big business, it is clear from this passage that for Watase the aim of wildlife protection is, above all else, to utilise 'the resources of the natural world' as an industry in order to generate profit.

We can assume then that the designation of natural monuments was instigated with this philosophy in mind. What kinds of animals were actually chosen and for what reasons? Let us now move on to the task of considering these questions in more concrete terms focusing our attention on whales.

The designation of the finless porpoise as a natural monument

The finless porpoise is a member of the family Phocoenidae (porpoises), adult specimens ranging from 1.2 to 1.9 metres in length. They live in coastal seas and major rivers from the Persian Gulf, through Southeast Asia to Japan. In Japan their distribution covers western Kyūshū, the Inland Sea and along the Japan Sea coast from Moji to Toyama Bay. On the Pacific coast they are found from the Kii Current to Sendai Bay (Carwardine 1995: 238–9; Kasuya 1994; Nihon Honyūrui Gakkai 1997: 142–3).

According to the ranking of the Mammalogical Society of Japan, the population of finless porpoises of Ōmura Bay is 'Endangered' and other established populations in Japanese coastal waters are 'Rare' (Nihon Honyūrui Gakkai 1997: 142–3).[4] And in the Inland Sea around Abajima (Hiroshima prefecture), the area within a 1,500 metre radius of Shirobana Iwa (White point rock) was designated a natural monument in 1930 as a finless porpoise migration site (Katō 1984: 47). In regard to the value of this species to the fishing industry, we have just a few reports of finless porpoises being taken for their oil for a period after World War II, or being captured for display in aquariums (Kasuya and Miyashita 1994: 630). We can conclude from this that there was very little commercial fishing that specifically targeted finless porpoises. Consequently, I think it would first be useful to trace the course of the designation of this species as a natural monument to get an idea of attitudes to conserving it.

A research report on the finless porpoise migration site was published in connection with its designation as a natural monument (Kaburagi 1932b). From this we can see that the designation was made under item 5 (See Table 3.1), but what was actually the reason the finless porpoise warranted designation? Let us see if an investigation of the contents of the report can enlighten us on this matter.

The author of the report, Kaburagi Tokio, gives two reasons for protecting the afore-mentioned finless porpoise migration site as a natural monument. One was that it was of scientific significance as the coasts of Honshū represent the northern limits of the distribution of the finless porpoise and the other was the fact that the fishermen of this area use a fishing technique known a *sunameri ajiro* (fishing with finless porpoises), so protecting the finless porpoise was 'essential for the fishing industry' (Kaburagi 1932b: 75). *Sunameri ajiro* is a fishing technique that exploits the fact that bream and sea bass gather near the seabed to feed on Japanese sand lance (*Ammodytes personatus*), a favoured food of the finless porpoise, which dives deep to avoid falling prey to the porpoises. The fishermen are able to catch bream and bass, usually difficult to catch in the winter months, by using live sand lance for bait and letting their boats ride with the tide around schools of migrating finless porpoises (Kanda 1981: 189; Shindō 1985; Katō 1984: 47).

So, clearly, we can see that the designation of the finless porpoise migration site was partly due to the fact that it was 'essential for the fishing industry.' Even so, we can think of the expression 'for the fishing industry' as it is used here as having three main connotations. The first of these is that the words 'significant' and 'unique,' which in a literal reading of the text of the schedule of categories for conservation as natural monuments, refer to animals that warrant designation, are applied in analogous fashion to the distinctive, unusual fishing technique of *sunameri ajiro* itself. Secondly, there is the sense that the finless porpoises are indispensable if the fishermen of this area are to use this technique, so they must be protected. In the same way, the designation, at the same time, of breeding sites of the black-tailed gull and the streaked shearwater as natural monuments, as the relevant reports indicated, was because of the importance of these species to fishing. The feeding behaviour of flocks of these seabirds told fishermen a lot about the location, composition and movement of shoals of fish (Kaburagi 1932c: 106–7; Kuzu 1932: 9–10, 16; Uchida 1925: 85–7, 96).[5] The third connotation includes the system of belief regarding these animals that evolved through the relationship between the finless porpoises and the fishermen. This becomes clear in the report on the sea surface flocking site, in Hiroshima prefecture, of the migratory Pacific diver (*Gavia pacifica*), also designated a natural monument at the same time (Kaburagi 1932a). According to the report, in the designated area of sea (the Ikari fishing grounds) a technique of fishing from rowboats under flocks of birds (*toritsuke sōchō gyogyō*) developed in the same way as the fishing with finless porpoises method, when Pacific divers began to replace the porpoises as the main predators of the sand lance. The fishing communities of this area would never capture or harm the Pacific divers, which they treated with affection and adoration as gods of the sea bream. Kaburagi, the author of the report, also claims, in what we might call folkloric research based on other similar instances in which the Pacific diver is revered, that 'there is something special in the spiritual relationship between fisherman and birds' (Kaburagi 1932a: 70) and concludes in his designation of the Ikari fishing grounds as a natural monument and his decision to protect the Pacific diver that 'this is appropriate not only for the sake of the fishing industry, but also on ideological grounds' (Kaburagi 1932a: 70). Kaburagi also states that the fishermen who employed the finless porpoise fishing technique treated the porpoises with the same loving care the fishermen of the Ikari fishing grounds did with the Pacific diver. He goes on to report how the fisherman had erected a shrine dedicated to the finless porpoise on Shirobana rock and how, on several occasions, fights had broken out between the local fisherman and fishermen from Ehime prefecture who had come into the vicinity of Abajima island in pursuit of finless porpoises (Kaburagi 1932b: 74).[6] So, we can assume that what he terms 'ideological grounds' was also a factor in the designation of the finless porpoise migration site.

It is clear from the above that protection of the fishing industry was a contributing factor in the designation of the finless porpoise migration site as a natural monument. In this regard, we can say it falls largely into the *conservation* category. Moreover, it is not the kind of conservation that leaves nature as it is, but conservation that calls for human intervention. However, it was not the case that the finless porpoises were being protected as a 'resource' that could continue to be exploited for profit. This was not the utilisation of a wild animal for major industrial production, but rather, what was being protected here was a relatively simple, subsistence level fishing activity and the relationship between the fishermen and the finless porpoises. Conversely, in this same period fox-farming in seabird breeding grounds, the provision of fishing port infrastructure, guano mining[7] and unbridled industrial development accompanying modernisation had placed seabirds in peril and led to the designation of breeding grounds as natural monuments. And the porpoise fishing technique itself fell into decline as the number of sand lances decreased in the area of the porpoise migration site due to the dredging of sand from the sea floor (Shindō 1985: 495).

This example of the finless porpoise shows that the designation of natural monuments is not just a matter of utilising wildlife for economic benefit and signals the need to question exactly what the concept of conservation entails. Let us now consider the direct protection of a species of wildlife for its economic benefit. As an example of just such a case, I propose to take a look at the grey whale.

The designation of the grey whale as a natural monument

Pre-World War II fishing industry regulations

The grey whale (1 family, 1 genus, 1 species) attains an adult length of from twelve to fourteen metres. Three populations have been identified; the American population in the North Atlantic and the American and Asian populations in the North Pacific.

It is thought that the these whales migrate from feeding grounds in the north to breeding grounds in the south, with the majority staying within a kilometre or two of the coast. The Asian population is thought to feed in the shallow waters of the central and northern Ohotsk Sea and in the winter months migrate south via the Korean Peninsula to the area around Hainan Island in southern China to breed (Ōsumi 1995; Carwardine 1995); [8] Nihon Honyūrui Gakkai 1997: 170–1).

The grey whale was eliminated from the North Atlantic in the eighteenth century and commercial whaling had reduced the American population to around two thousand whales by the beginning of the twentieth century. As a result of a ban on the commercial exploitation of grey whales introduced in 1946 the numbers have been restored to a point where today the population is estimated at around 23,000. In contrast, the Asian population, according to the ranking of the Mammalogical Society of Japan, is now 'Endangered.' It is assumed that the cause of this decline was large-scale fishing of this species by the Japanese whaling industry in the coastal waters of the Korean peninsula under Japanese colonial rule (Nihon Honyūrui Gakkai 1997: 170–1; Ōsumi 1995: 516–18). Figure 3.1 shows variations in the numbers and species of whales taken off the Korean coast during the colonial period. We can see from the chart that the numbers of grey whales taken gradually decreased from around two hundred a year in the early days when it vied with the fin whale as the most important target of Japanese whaling on the Korean coast, to the point where not a single grey whale was taken after 1934. We cannot help but wonder what measures might have been taken at the time (before World War II) to protect the dwindling population of grey whales. Does the fact, shown in Figure 3.1, that no grey whales were taken after 1934 reflect any kind of regulation policy? When we look at the

Figure 3.1: Whales taken in Korea under colonial rule

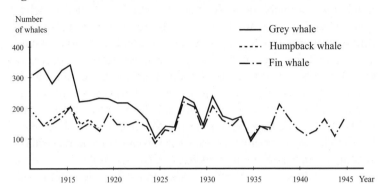

Note:

1. There is insufficient data to complete the graph for some species where the numbers taken were very small; often no more than one or two whales per season.

2. The numbers of blue, sei, sperm and right whales taken were too small to show on the graph.

Source: (Park 1995: 526–7) (Original source: Nihon Hogei Kyōkai, *Hogei tōkeibo*, for the years 1911–25 and 1940–44); Nōrin Daijin Kanbō Tōkeika 1927–30, 1932–6, 1938–40. 'Dai sanjūrokuji Nōrinshō tōkeihyō.'

contemporary statistics of catches of grey whale in 'Japan,' we see that almost all were taken in the waters of the Japanese colony, Chōsen (Maeda and Teraoka 1952: 106–7; Park 1995: 526–7).

So in the following discussion we consider the whale protection system on the ruling side, in the so-called 'home islands' (*naichi*), on the one hand and that in Chōsen under colonial rule on the other.[9] First let us take a look at regulation of the fishing industry.[10] According to the Regulations for the Management of Whaling (*Geiryō torishimari kisoku*) issued in 1909, Norwegian-style whaling came to require the approval from the Minister of Agriculture and Commerce. And if he approved the request, the Minister would then decide the species of whales to be taken, the period of the licence, the area to be fished and the number of ships to be used. The regulations gave the minister the authority to ban or limit whaling and to mark designated ships (*Kanpō* (Official gazette) 7899, 21 October 1909). Accordingly, in 1909 the number of whale catchers was limited to 30 and this figure was reduced to 25 vessels in 1934 (*Kanpō* (Official gazette) 7899, 21 October 1909) (*Kanpō* (Official gazette), 2245, 27 June 1934) (See also Note 2 in Chapter 5). Further, it seems that the limit on the number of whale catchers introduced in 1909, was based on the number of ships engaged in whaling at that time.

From the viewpoint of the protection of whale breeding stocks this number was already too high. When the number of whale catchers was reduced further in 1934 it was pointed out that the figure of thirty ships was too large, given the migratory patterns of whales and the fact that the catch of baleen whales had been decreasing annually (Shigeta 1962a: 16–17). Later, in 1934, Regulations for the Management of Factory Ship Type Fisheries were introduced to control the industry (*Kanpō* (Official Gazette) No, 2269, 25 July 1934). And in the Regulations for the Management of Whaling of 1938 the taking of young whales, suckling calves or mother whales accompanying their young was banned (*Kanpō* (Official Gazette) No, 3427, 8 June 1938). Further, limits on the body length (or bans on the taking of undersized whales) were imposed on blue, fin, humpback, sei and sperm whales, and the two sets of Regulations for the Management of Whaling, cited above, with a few variations in regard to body length, were appended to the Regulations for the Management of Factory Ship Type Fisheries, along with a ban on taking grey or right whales outside that area of the North Pacific Ocean above the latitude of twenty degrees north.

It seems safe to assume that these regulations were incorporated into domestic law as an indication of Japan's respect for the aims of the International Whaling Agreement to protect whale stocks and stabilise the price of whale oil, that was signed in London in 1937 (See Chapter 1) (Baba 1942: 308–10; Ōmura, Matsuura Yoshio *et al*. 1942: 307, 309; Shigeta 1962a: 19; 1962d: 16).

In addition, under the Government-General of Chōsen,[11] as a result of the Fisheries Ordinance issued in 1911, whaling required the permission of the Governor-General (Hanguk Hakmun Munhyeon Yeon'gujang (Korean Institute for Research into Academic Sources) ed. 1985–8).[12] Further, the number of whale catchers was restricted to ten vessels under the 1917 amendment of the Rules of Enforcement of the Fisheries Ordinance, though this number was raised to twelve in 1922 (*Chōsen Sōtokufu Kanpō* (Official gazette of the Government-General of Chōsen) No. 1572, 1 November 1917, No. 3092, 1 December 1922). Later, under the Rules for the Enforcement of the Chōsen Fisheries Ordinance (an attachment to the Chōsen Fisheries Ordinance that replaced the Fisheries Ordinance in 1929), the designation of whaling zones and the quotas on the number of whale catchers in these designated areas came under the discretion of the Governor-General of Chosen. The number of whale catchers permitted 'for the coastal and neighbouring offshore waters of Chōsen' was maintained at twelve vessels (*Chōsen Sōtokufu Kanpō*, extra edition, 10 December 1929). But in March 1944 catcher boats to take minke whales were exempted from the quota system and in October of the same year the limitations on whaling zones and the quotas on the number of whale catchers were abolished altogether. We can assume that these measures were in response to the military mobilisation of whaling vessels and food shortages resulting from the war (*Chōsen Sōtokufu Kanpō*, No. 5136, 20 March 1944; No. 5315, 21 October 1944).[13] In addition, since the enactment of the Regulations for the Management of Fisheries in 1911, there had been a ban on whaling from 1 May until the end of September, and even outside this period, it was prohibited to take whale calves or their accompanying parents (*Chōsen Sōtokufu Kanpō*, No. 227, 3 June 1911).[14]

But in the 1921 amendment to the Regulations for the Management of Fisheries, these restrictions (the five month no-whaling period and the ban on taking calves or adults accompanied by calves) were lifted (*Chōsen Sōtokufu Kanpō*, No. 2611, 27 April 1921) and did not reappear when the Management Regulations for the Protection of Chōsen Fisheries replaced the Regulations for the Management of Fisheries in 1929. Moreover, in the new regulations of 1929, and in subsequent amendments, no mention was made of protection of specific species of whale (*Chōsen Sōtokufu Kanpō*, extra edition, 10 December 1929).[15]

It is clear from our investigation so far that in practical terms the Asian population of grey whales gained no protection from the pre-war rules governing fisheries. And considering the protection of whales overall, we have seen how, over the years, a progressive tightening of the regulations for whaling in the Japanese home islands (*naichi*) contrasted with a relaxing of restrictions in Japan's colony, Chōsen.

The road to designation as a natural monument

Let us now consider how the grey whale came to be a natural monument. Once again we need to turn our attention to the categories in the schedule in Figure. 3.1. Interestingly, the grey whale is given as an example for category ii. in the schedule. Of course, as the preamble states, the fact that an animal is given as an example does not imply that it will be immediately designated a natural monument. Nevertheless, I think perhaps we can safely say that by 1919 when the schedule was drawn up, the grey whale was considered a worthy candidate for designation as a natural monument. Further, a paper appeared in 1926 arguing for the designation of the grey whale (Tago 1926). The author of the paper, Tago Katsuya, relates how the grey whale is found only in the North Pacific and that those that formerly migrated along the west coast of the United States (the American population) are believed to be virtually extinct.

He goes on to point out that even in the seas surrounding Japan there has been a marked decline in catches of grey whales (the Asian population) in recent years. He claims that not only is the grey whale of great scientific interest, but is also of considerable economic value to the fishing industry as it is one of the few species of whale to venture into eastern waters in the winter months. Tago concludes (Tago 1926: 1–2, 14–15),

> However, fortunately, it was recognised that in order to prevent overfishing we needed to consolidate the various companies formerly licensed to engage in whaling in Chōsen into a single company. That single company, the Tōyō Whaling Company, is now operating under a policy of strict self-restraint in line with the wishes of the government-general of Chōsen to ensure it does not overfish. In particular, given the attention being paid recently to the protection of grey whales, and if there is no change in the current policy of the government-general, I do not believe there is any immediate danger of grey whales becoming extinct, but if we do not adopt an appropriate protection policy I am sure the day will come when we regret our failure to do so (Tago 1926: 15).

However, it was 1909 when several whaling companies combined to form the Tōyō Whaling Company, which, for practical purposes, started whaling under its monopoly around the Korean peninsula in 1910 (Park 1995: 282) (See also Chapter 1). Table 3.1 indicates that it was the period after this, particularly from the 1920s that saw the beginning of the rapid decline in grey whales. And, as we have just seen, from the 1920s the government-general of Chōsen moved down the path of relaxing its fishing restriction in regard to whales and there was no mention made of the grey whale. The possibility remains that Tōyō whaling was exercising self-restraint under some kind of rules it had devised itself, but Tago's belief that extinction could be avoided by following the government-general's current policy proved to be incorrect, in the sense that grey whale numbers continued to decrease. Moreover, in spite of this serious situation, the grey whale was not designated a natural monument within the Japanese home islands at that time, nor is it so designated even in Japan today. On the other hand, under the Ordinance for the Conservation of Treasures, Ancient Sites, Scenic Places and Natural Monuments in Chōsen, issued by the government-general of Chōsen in 1933, replacing the Regulations for the Conservation of Historical Sites and Relics of 1916, the government-general of Chōsen began to designate natural monuments and provide for their protection (Chōsen Sōtokufu 1934: 1–6).

In Table 3.2, as in Table 3.1, I have extracted the section on animals from the schedule as a whole.[16] Here too, in much the same way as in Table 3.1, the grey whale is given as an example of category ii. And in 1942 the Ulsan grey whale migration site along the coasts of Gangwondo, Gyeongsangbukdo and Gyeongsangnamdo provinces was designated a natural monument (Designation No. 126) (*Chōsen Sōtokufu Kanpō*, No. 4612, 15 June 1942).[17] The designation meant that the grey whale became eligible for systematic protection under articles five and six of the Ordinance for the Conservation of Treasures, Ancient Sites, Scenic Places and Natural Monuments in Chōsen, which state respectively, 'any activity that may change the *status quo* of the designated item or influence its conservation must have the approval of the Governor-General,' and 'when he deems it necessary for the conservation of the designated item, the Governor-General … may ban or limit a specific activity, or order the provision of necessary infrastructure' (*Chōsen Sōtokufu Kanpō*, extra edition, 8 August 1933). But we should probably question whether the designation functioned according to the letter of the law in the context of World War II food shortages and the consequent relaxation of controls on whaling.

From the above we can assume that even the protection afforded under designation as a natural monument had been shelved. True, the grey whale was declared a natural monument in Chōsen, but considered in the broader contemporary framework of greater Japan, including its colonies, this was twenty years after the period when debate over its designation began. Moreover, it seems very likely that the fact that Figure 3.1 shows that no grey whales were taken after 1934, notwithstanding the possibility that the whaling company was practising self-restraint, was because of the decline in the population of grey whales, rather than the result of any administrative measures, including those imposed by the fishing industry.

Why was it, then, that the regulations pertaining to grey whales were ignored to a point where the species was driven to the verge of extinction?

Table 3.2. Schedule of categories for the Conservation of Treasures, Ancient Sites, Scenic Places and Natural Monuments in Chōsen

(Preamble omitted)

Natural Monuments

The following categories have been approved for conservation as natural monuments

I. Animals

   i. Those significant species that are currently found in Chōsen and have not yet been discovered elsewhere in the world (e.g. the white-bellied woodpecker (*Dryocopus javensis*) which is only found in Chōsen and on Tsushima Island).

  ii. Those species that until comparatively recently lived in other parts of the world, but have gradually decreased in number to the point where they are now found only in the home islands (*naichi*) and in Chōsen (e.g. the grey whale (*Eschrichtius robustus*) of the Japan Sea).

 iii. Those species that formerly lived in Chōsen in considerable numbers but have recently decreased to a point where they are facing extinction (e.g. the oriental stork (*Ciconia boyciana*), the black stork (*Ciconia nigra*), the spoonbill (*Platalea leucorodia*), the crested ibis (*Nipponia Nippon*), the great egret (*Ardea alba*), the intermediate egret (*Egretta intermedia*), the great bustard (*Otis tarda*) and the sable (*Martes zibellina*) of Hamgyeongbukdo, etc.)

  iv. Those species, the conservation of which is desirable because, though not unique to Chōsen, they are of considerable significance in East Asia (e.g. the Chinese water deer (*Hydropotes inermis*)).

   v. The habitats of significant animals in Chōsen.

  vi. Breeding grounds or migratory sites of significant animals (e.g. the migratory sites of cranes in Jincheon in Chungcheong, Baecheon in Hwanghaedo and Goheung in Jeollanamdo, the oriental storkbreeding ground Ganghwado in Gyeonggido and the breeding grounds of seabirds like the guillemot (*Uria aalge*) and the ancient murrelet (*Synthliboramphus antiquus*) in the Japan Sea).

 vii. Fossil remains of various giant mammals, such as the elephant, rhinoceros, sabre-toothed tiger and deer, and other significant animals found in Chōsen and the sites where they were discovered.

viii. Animals and groups of animal species as a whole unique to habitats such as, mountain, plain, marshland, forest, wetland and lake, seashore, river and sea, island, cave etc.

  ix. Domesticated animals unique to Chōsen (e.g. the Chōsen dog)

   x. Significant animal habitats on adjacent islands.

(Following omitted)

Note: The categories given above are interim examples provided for reference for the reports from the various district administrations, or to illustrate legal terminology, accompanying the issue of the Conservation Ordinance. They are to be amended at an appropriate time. Further, the examples given are for reference only and will not necessarily be designated under the Conservation Ordinance.

Source: (Chōsen Sōtokufu 1934: 7–23)

Much of the detail of the debate over fisheries regulation and the designation of natural monuments remains unclear. I have not been able to access minutes of meetings and there is a high possibility that these no longer exist. Nevertheless, I think we can assume that events played out along the following lines. Within the extended territory of Japan at the time, the grey whale was an extremely important *resource* for the whaling company. This was particularly so, given that the whales were principally taken in Japan's colonial waters. Further, since the grey whale was used for economic production, any regulation of that resource meant that the whaling company had to voluntarily slow down or reduce the scale of its production. This leads us to suspect that perhaps it was only when the grey whale was facing extinction and was no longer a viable *resource* that protection measures were finally implemented.

Before bringing this section to a close let me say a little about regulations applied to the grey whale after World War II. According to the statistics, Japan has caught no grey whales since 1946 (Nōrin Suisanshō Tōkei Jōhōbu: Nōrin Tōkei Kenkyūkai 1979: 258–63). In addition, the taking of grey whales was prohibited under the then current regulations of the International Convention for the Regulation of Whaling, which Japan formally joined in 1951 (Maeda and Teraoka 1952: 244–51). Further, although around ten grey whales a year were taken until the end of the 1960s in Korea (the Republic of Korea), we can assume that none have been taken since. Moreover, the Ulsan grey whale migration site along the coasts of Gangwondo, Gyeongsangbukdo and Gyeongsangnamdo provinces has been designated a Korean natural monument (Park 1995: 360, 398–400, 527).

Summary

In this chapter we have considered the process and substance of protection of the finless porpoise and the grey whale in the context of the logic underlying the designation of natural monuments. Let me recapitulate what our investigation has brought to light in regard to the use of wildlife for industrial production.

Together with their academic significance and value to science, the reasons given for requesting the designation of the finless porpoise and the grey whale as natural monuments included their importance to the fishing industry. In other words, their designation as natural monuments can be seen in a sense as protection for the fishing industry.

Seen in this light, the designation appears to be an appropriate application of Watase's idea of the protection of wildlife so it can be utilised by industry. Closer consideration of the actual developments, however, revealed that it was not as simple as this. In addition, there were substantial differences in the circumstances surrounding the finless porpoise and the grey whale. Certainly, the grey whale was thought to be worthy of protection because of the economic benefit it provided. But in fact, the whales did not become candidates for systematic protection until there were no longer enough left to catch. Rather, we can speculate that it was because they were used as a *resource* by the powerful whaling industry that grey whales were driven to the verge of extinction.

We can suggest reasons for this mismatch between the theory and practice. One is the fact that there is a fundamental incompatibility between the economic activity of using wild animals for production and the protection of those same animals. The idea of protection of the *resource* may sound plausible, but in reality industry cannot allow the brake on production that this implies. Another reason, in the case of the grey whale, may reside in the fact that whaling for this species occurred mainly in colonial waters. That is to say, under the expansionist ideology of colonial rule the *resources* of the colonies were there to be exploited to the limit. When they were exhausted new sources of supply were sought elsewhere.

In Chapter 5 I shall be considering these points in a little more detail, but here I would like to add a few remarks regarding the current debate over protection of wildlife. I think the arguments we hear today, for example that we should (re-)combine the system of wildlife protection with the market economy or that we should direct the profits made from the utilisation of wildlife into protection programs, are far too simplistic. To take the example of the Hokkaidō *shika* deer; in the past, the numbers of these animals had been greatly reduced through indiscriminate hunting and heavy snowfalls, but from the 1960s the felling of native forests and the planting of commercial timber, the clearing of land for pasture and the construction of forest roads, provided food sources and safe habitats for the deer. These developments were the primary cause behind the rapid increase in the numbers of deer, which resulted in damage to agriculture and the forestry industry in excess of fifty billion yen (approximately forty-five million US dollars) a year. In March 1998 the 'Conservation and Management Plan for *Shika* Deer in Eastern Hokkaidō' was introduced as a strategy to deal with this problem (Hokkaidō Kankyōseikatsubu Kankyōshitsu

Shizenkankyōka 1998; Kaji 1999; Ōtaishi and Honma1998: 1–16).[18] Put simply, the system worked something like this. The population of deer at the time was estimated at around 120,000 (fixed population index 100). Taking this number as its baseline, the plan provided three levels of management, the 'irruption threshold' (index 50), the 'optimal level' (index 25) and the 'critical threshold' (index 5). Depending on the population of deer within these levels, a management strategy was chosen from the following four options; 'emergency culling,' 'gradual population reductions,' 'gradual population increases' and 'hunting bans.' Further, a large-scale cull of *shika* deer, concentrating on does, was put in place for three years from 1998 to reduce the population to the 'irruption threshold.' After that, the aim was to maintain the population at the 'optimal level.' Actually, in the 1998 season a total of 73,000 *shika* deer, comprising 35,000 bucks and 38,000 does, was culled in Eastern Hokkaidō. A total of around 50,000 *shika* deer have been hunted each year since then and in the course of this trend there have been moves to make 'effective use' (*yūkō katsuyō*) of the deer taken, such as, for instance, through the circulation of venison into the marketplace (Ōtaishi and Honma1998). But movement in this direction is problematical. Currently, virtually no market has been established for venison. But let us assume for the moment that through various methods of promotion, along the lines that have been implemented for whale meat (See Chapter 4), the consumption of *shika* deer meat has gained popularity in the market. Because *shika* deer are vulnerable to heavy snowfalls, under the management plan introduced above, in the year following a very severe winter, 'consideration is given to the need for a prohibition on hunting, taking into account the population index of the previous year' (Hokkaidō Kankyōseikatsubu Kankyōshitsu Shizenkankyōka 1998: 5).

In this way the brake of the hunting bans is applied. But in so much as the *shika* deer has become a *resource* to be used for economic production, it follows that the industry must employ a range of people in the various stages of production, like processing, transport and so on. So, is it really possible to deprive these people of their livelihood through the imposition of a hunting ban? Moreover, if we are to believe the figures put out by those who advocate 'effective use' of the *shika* deer,[19] we could well point to the possibility of low-priced foreign venison flooding the market they had established. If that were to happen, and the *shika* deer they had shot could not be sold, they would become less enthusiastic about shooting them. And if, for the sake of argument, we were to assume that implementation of the plan of management described above depended on the profits directed to it from the 'effective use' of the *shika* deer, the whole project of controlling the deer population through hunting would be deprived of funds and thrown into disarray.

In short, combining wildlife protection with economic production inevitably means that activities involving the protection of wildlife end up being swamped by the forces of the market economy. In other words, the original goals of appropriate wildlife protection and management are overturned when the focus becomes fixed on wildlife as a *resource* to be used for economic gain. Consequently, in order not to arrive at the unhappy situation we observed with the grey whale, even though emergency measures involving mass culling may sometimes be inevitable, I think it is preferable to limit the use of the *shika* deer shot by hunters, (the majority are killed under pest extermination programs and buried as general waste) (Ōtaishi and Honma1998: 14, 129–31), to personal family consumption.

Now, let us move on to consider the finless porpoise in the light of the arguments advanced above concerning the grey whale and the Hokkaidō *shika* deer.

One aspect in the designation of the finless porpoise was to protect, not a major industry, but small subsistence fishing by people who had a deep connection with the finless porpoise. So, the porpoises were not protected to ensure the continuation of what could be gained from treating them as a resource. Rather, we can consider that they were protected in order to *prevent* them from becoming a resource. Further, we saw how, in the case of the finless porpoise, the reasons for designation as a natural monument included 'ideological' considerations, reflecting the existence of a system of belief that had grown up around the relationship between the fishermen and the porpoises.

These reasons for protecting the finless porpoise are quite different in nature from those applied in the protection of wild animals for economic gain or in fixating on wildlife as a *resource*. It is difficult to see the designation of the finless porpoise resulting in the overfishing of bream and bass caught by the porpoise fishing technique. Neither can the decline in this fishing method be linked to the overfishing of bream and bass, let alone finless porpoises. So, this is an interesting example of a relationship between wildlife and human beings, in the sense that it shows how wildlife protection can be appropriately combined with human lifestyle and subsistence economic activity within a *conservation* approach. Nevertheless, you may say, it is problematical to idealise this case as 'environmentally friendly' simply because it involves the folkloric element of popular belief. On the contrary, I think it has become necessary in the debate over wildlife protection today to include consideration of the kinds of issues that surface in this case, including meta-level considerations, like the logic behind idealising this example.

4 The Promotion of Whale Meat in Early Modern Japan

Introduction

In this chapter I seek to clarify how and to what extent the consumption of whale meat came to spread throughout Japan in the early twentieth century. In doing so, I would like to analyse the issue from the following perspective. Given particular socio-historical conditions, even food habits, in other words, the everyday activity of choosing ingredients, processing, then eating them, is not 'a fixed, unchanging activity among people who are regarded as constituting a group.' Rather, we can consider it as taking its specific form through an on-going process. The task of this chapter is to consider that process.

Incidentally, what I am calling 'socio-historical conditions' does not simply imply advancement in the techniques of distribution and preservation. Of course, advances in technology are no doubt necessary for a new food to become accepted into the daily diet, but it does not necessarily follow that if there are advances in technology, people will start eating a particular food. In this chapter, rather than considering advances in technology, I shall be focusing on changes in policies and conditions that penetrated the most intimate corners of the lives of the people of the time, and on what might be termed 'cultural influences,' which helped fashion people's tastes.

In considering these shifts and changes, we must not overlook the fact that an aspect of power comes into the equation, in the form of 'food hegemony' (Barsh 2001: 150–61). That is to say, that, as in the Roman Empire or under European Imperialism or in Western control of financial and food commodity markets, food production and patterns of consumption in colonies or areas under the control of dominant powers is changed to suit the requirements of the ruling elites.

Bearing these points in mind, first let us see what historical documents can tell us about how the habit of eating whale meat spread throughout Japan. Then, I would like to clarify the extent of whale meat consumption by analysing the results of a survey taken around the time of the outbreak of World War II.

Conditions behind the promotion of whale meat consumption

Today, particularly in the context of the so-called 'whaling issue,' our attention is repeatedly drawn to the discourse that the Japanese have been eating whales since time immemorial. Recent research in particular has shown, however, that this narrative is problematical. For example, Morita Katsuaki, addresses the argument that has surfaced in the media since the whaling moratorium (initiated at the end of 1987), that Japan is a 'whale-eating culture' and the claim that eating whale meat is part of the traditional food culture of 'the Japanese race.' He points out distortions in the data supporting this argument as follows:

> Certainly, whale meat consumption has a long history, but it was not until after World War II that whale meat came to be eaten on a daily basis over the nation as a whole. Further, I hasten to point out the danger in linking the consumption of whale meat with the vague and highly political term, 'the Japanese race' (Morita 1994: 414–15).

While not denying the importance of this criticism, I would like to know, if indeed the eating of whale meat was so wide spread, why, and to what extent, this was so; questions not actually clarified in Morita's work. In the following, I seek to drill down on these matters by pursuing the debate through historical sources. Before doing so, however, I would like to point out that whale meat can be broadly divided into; *akaniku* ('red meat' – brisket, the meat of the torso, usually distributed raw), *shironiku* or *shirotemono* ('white meat' – the fatty layer (so-called skin), the ventral pleats (the corrugated area running from the lower jaw to the end of the belly), the tail flukes, the pectoral fins and so on, usually sold salted).

To get to the main point, it is known that since ancient times use was made of the carcasses of whales drifting ashore (*nagare kujira*) or stranded on reefs or beaches (*yori kujira*) (Fukumoto 1978: 25, 42–7). Of course, we can assume that the meat of these beached whales was eaten. And no doubt whale meat was also eaten in and around present-day Nagasaki–Saga–Fukuoka, Kōchi and Wakayama, where net whaling was introduced at the end of the seventeenth century. But it seems that these were extremely

localised occurrences. Actually, until the end of the nineteenth
century the areas where whale meat was eaten and the volume
consumed gradually decreased, from a peak in northern Kyūshū
corresponding to those areas where net whaling was practised
(Nagasaki–Saga–Fukuoka), as you moved east into the Kansai
area and further on into eastern Japan. *Akaniku* in particular, was
apparently hardly eaten at all east of Nagoya in 1912 (Andō 1912a:
16; Maeda and Teraoka Yoshirō 1952: 170–7).

The momentum for change in this pattern of whale meat consumption was provided when Norwegian-style whaling was
introduced in 1897 and, later, in 1909, with the birth of the monopoly
whaling company, Tōyō Whaling, through the amalgamation of the
numerous whaling companies that had sprung up after the Russo-
Japanese war (See Chapter 1). With these developments whaling
become a very large industry. And the whaling company strove to
increase its profits, not only by using modern technology to catch
more whales, but also by creating more markets for the products it
obtained from them.

In the first half of the nineteenth century, in the Kansai area
(Ōsaka–Kōbe), nowadays one of the major areas for consumption of
whale meat, the Ōsaka market, handled *kawakujira* (probably white
meat) and *mikujira* (probably red meat) from Kii (Wakayama) and
Tosa (Kōchi) (Izukawa 1943: 291, (1973c: 311)). In addition, since we
have a record from the latter half of the nineteenth century of the red
meat of a whale caught in Tosa being transported to the Ōsaka–Kōbe
area (Izukawa 1943: 392–8, (1973c: 311); Yoshioka 1938: 23–32),
we can assume that red meat was eaten before the introduction
of Norwegian-style whaling in Kansai (Ōsaka–Kōbe).[1] But prior
to the advent of Norwegian-style whaling, the eating of whale
meat is considered to have been restricted to the urban samurai
and townsmen (merchant) classes (Izukawa 1943: 162–3, 543–4
(1973c: 182–3, 575–6)). Moreover, the number of whales caught in
a year under the net-whaling method, if we take the Tsuro group of
Tosa for example, according to the *Tsuro hogeishi* (An account of
Tsuro whaling) averaged 20.6 whales per year between 1693 and
1712, 21.8 between 1849 and 1865, 16.8 in the period 1874–90 and
16.5 from 1891 to 1896 (Izukawa 1943: 1–11, (1973c: 21–31); Tsuro
Hogei K.K. 1902: leaf 106 (front) – leaf 109 (front)).[2] These figures
are extremely low compared to the number of whales taken with
Norwegian whaling technology. Consequently, I think it is safe to
assume that over the Kansai (Ōsaka–Kōbe) area as a whole, the
consumption of whale red meat was not particularly high. In actual

fact, there are accounts to the effect that it was not until around the time of the introduction of Norwegian-style whaling that red meat became readily available in Kansai (Ōsaka–Kōbe). Even then, as we can imagine from an account of a whale meat vendor having to pay a cleaning fee to a customer whose clothes had been stained by the blood dripping from his *akaniku* purchase as he made his way home, the product was difficult to sell (Maeda and Teraoka Yoshirō 1952: 171, 174–5). But the situation changed dramatically when Yamada Tōsaku, who was in charge of sales for Tōyō Whaling, formed the Maruichi Shōkai company around 1909 (the firm changed its name to Isana Shōkai in 1915),[3] to sell whale meat. As a result of his efforts to expand the whale meat market, according to a contemporary account, by around 1912, 'the vigorous growth in whale meat sales in the Ōsaka–Kōbe area was astonishing and increasing at the rate of fifty per cent per year' (Andō 1912a: 16).

Tōyō Whaling published its *Honpō no Noeruēshiki hogeishi* (A history of Norwegian-style whaling in Japan) in 1910, both as a company history and to introduce the whaling industry and promote its products. With an eye to broadening the market for whale meat, the book includes a constituent analysis of whale meat and recipes illustrating how the meat is used in Kyūshū. Further, and one can only assume that this was a suggestion from the Tokyo branch office of Tōyō Whaling, there is a section headed 'Whale meat recipes for Eastern Japan' (*Geiniku no Azuma ryōri*) (Tōyō Hogei K.K. 1910: 38–59, 125–38). In addition, in 1919, through the good offices of the Fisheries Bureau of the Ministry of Agriculture and Commerce, a whale meat bargain sale was held in Tokyo. Tōyō Whaling, abandoning any thought of profit, displayed its whale meat in thirty-three riverside fish markets around Nihonbashi, three Tokyo City public markets and six Tokyo Prefectural public markets ('Shokuryōhin no hippaku to geiniku shokuyō' (Scarcity of foodstuffs and the consumption of whale meat) 1919: 35) 'Geiniku o uru' (Whale meat to go on sale) (1919).[4] And from around 1921, *akaniku* came to be systematically sold in Tokyo from its stronghold at the Hoteiya store (the forerunner of the Isetan department store in Shinjuku) (Maeda and Teraoka Yoshirō 1952: 171, 175). A 'Whale Exhibition,' organised by the Ministry of Agriculture and Forestry, Tōyō Whaling and others involved in the whaling industry, was held at Hoteiya in 1932 to advertise whale meat (Nakatani 1932).

In addition to these efforts by the whaling company to enlarge the market for its products after the introduction of Norwegian-style whaling, I think the fact that modern Japan had been involved in a number of wars had some bearing on the diffusion of the habit of eating whale meat. Actually, war and the whaling industry are closely entwined, not only in regard to the spread of whale meat consumption. I explained in Chapter 1 how the initial introduction and establishment of Norwegian-style whaling was abetted by the fact that Tōyō Whaling's predecessors gained the whaling ships and whaling concessions on the Korean peninsula that had been confiscated from Russian whaling companies at the time of the Russo-Japanese War (1904–5).

Maeda Keijirō and Teraoka Yoshirō maintain that it was the influence of the first and second world wars that led to the large scale consumption of whale meat and established its importance as a food source.

Whale meat was welcomed as a very cheap alternative at a time when there was a general downturn in household purchasing power, during the depression that hit Japan in the wake of the First World War, and rapidly spread through the market. There was a further leap in whale meat consumption, for military provisions and in response to food shortages from the time of the so-called 'Manchurian Incident' of 1931, through the Pacific War and on into the post-war period (Maeda and Teraoka Yoshirō 1952: 170–3). Against this historical background, in the pre-war period, factory-ship whaling in the Antarctic Ocean, which used a mothership as a base for flensing and processing the whale carcass at sea, began to produce meat, fertilizer, leather and fibre from those parts of the whale that were not a source of whale oil and which, in the early days of the industry, had been discarded into the sea. Nevertheless, it should not be forgotten that whale meat production was essentially a by-product of secondary importance to the whale oil, which could be exported to Europe to earn foreign currency and was important for military purposes (See Chapter 1).

Here, in regard to the influence of war on growth in the popularity of whale meat, it is necessary to say a little about the whale meat canning industry. Originally, in Japan it was the need to supply military provisions during the Sino-Japanese War (1894–5) that provided the impetus that propelled canned food in general from its initial trial stage into a fully-fledged processing industry. After that, the navy established designated plants in a dozen or so 'strategic locations' throughout the country to ensure that there would be no shortage in the supply of canned foods in peacetime or during emergencies. In addition, the army set up the Central Army Provisions Depot (*rikugun chūō ryōmatsushō*) in 1897 and in 1900 introduced the latest canning machines from Germany to complete its preparedness for any eventuality by manufacturing canned foods, mainly beef. Later, during the Russo-Japanese War, the army requested the Fisheries Bureau of the Ministry of Agriculture and Commerce to provide, under the supervision of the Bureau, a stable supply of low cost, high quality canned seafood for military use (Asahina 1915: (1997) 37–50, 66–70, 192–205). In addition to the background described above, it seems the canning of whale meat started from around this time as a solution to the problem of handling the *akaniku* (red meat) from spring through to the end of summer (Andō 1913: 14–15; Maeda and Teraoka Yoshirō 1952: 173).[5]

An 1897 list of canners and canneries, for example, indicates that three (one in Hyōgo and two in Kōchi) were manufacturing a product labelled 'whale' or 'whale meat.' Further, at the time of the Russo-Japanese War, another list of canneries producing canned seafood for military use, includes three canners (one each in Ishikawa, Yamaguchi and Kōchi; the Yamaguchi canner was Tōyō Fisheries, the precursor to Tōyō Whaling) that produced products marked as 'whale.' But, seen in relation to the list as a whole (three out of one hundred and fifty-seven in the former case and three out of one hundred and fourteen in the latter), whale meat accounts for a tiny proportion of the total canned foods produced (Asahina 1915: (1997) 50–64, 199–205). Moreover, an account dated 1913 to the effect that 'we make very little money from the various processed products (apart from *akaniku*, salted *shironiku*, fertiliser and whalebone),' suggests that at this stage the canned whale meat industry was fairly insignificant. After that, however, we can assume that there was a steady increase in canned whale meat production with the growth of whaling year after year in Tōhoku and Hokkaidō, together with the large-scale production for military use during the Manchurian Incident and the Pacific War. Incidentally, by 1952 the Tōhoku region (northern Honshū) boasted the country's highest consumption of canned whale meat, which, if we are to believe a contemporary account, was regarded there as indispensable for festivals and other celebrations (Maeda and Teraoka Yoshirō 1952: 175).[6]

To sum up the argument so far, we can surmise that the eating of whale meat on a limited scale began to spread over the whole country from the end of the nineteenth century, as a result of vigorous promotion by the whaling industry, which had become a major economic force with the introduction of Norwegian whaling technology. What is more, we saw how whale meat consumption was closely tied to the contemporary behaviour of the Japanese State, for strategic purposes and the waging of war. As a result of these actions, according to one theory, *shironiku* (white meat), which had been the most commonly eaten whale meat, had been overtaken by *akaniku* (red meat) by the end of World War II. Further, it is claimed that even the *akaniku* of the sperm whale, usually regarded as rather unpalatable, came to be a widely consumed food item (Maeda and Teraoka Yoshirō 1952: 173).

But just how widely was whale meat actually eaten in Japan at that time? To elucidate this question let us look at the following concrete figures.

Analysis of a survey on whale meat consumption

The source material

My source for this section is the data from a survey made by Izukawa Asakichi, presumably in 1941 (Izukawa 1942). To the best of my knowledge, this is the only extant survey of the dissemination of whale meat consumption in Japan before World War II. Further, apart from a few brief comments on the results of the survey by Izukawa himself, and subsequently by Yamashita Shōto (Yamashita 2004a: 210), I think we can say that there has been no other detailed analysis of the survey, that considers the statistical data and seeks to interpret the implications of each of the written responses. For this reason, I think the survey is both appropriate and necessary for the task of this chapter, namely, to examine the extent of whale meat consumption in modern Japan.

The survey, requiring an open-ended written response to specific questions, was conducted in the two metropolitan regions (*fu*) and thirteen prefectures (*ken*) of the Kinki and Chūbu areas of central and western Honshū. To ensure a good rate of responses, the questions were printed on a pre-paid return postcard and most were distributed through primary schools covering the whole of the area to be surveyed. The number of surveys distributed and the number of responses are summarised in Table 4.1. There is some variation in the rate of returns from one prefecture to another, but on the whole the response rate is not unreasonable for a postal survey. Although nine hundred surveys were sent out, two of these were 'imperfectly printed' (Izukawa 1942: 115 (1973a: 411)).

Table 4.1: Number of responses to the survey

	Sent	Returns	Rate %
Kyoto	79	21	27
Osaka	44	17	39
Wakayama	42	17	40
Mie	60	21	35
Nara	41	10	24
Shiga	43	11	26
Fukui	39	9	23
Gifu	87	41	47
Aichi	40	13	33
Shizuoka	52	10	19
Nagano	121	32	26
Yamanashi	59	17	29
Ishikawa	60	12	20
Toyama	67	23	34
Niigata	64	25	39
Totals	898	279	31

Note: Percentages rounded to the nearest whole number

Source: (Izukawa 1942: 113–145 (1973a 407–41))

The gist of the six questions, which I have abbreviated somewhat, was as follows; 1. How long have you been eating whale meat? 2. Is that meat blubber or *akaniku* (red meat)? 3. Throughout the year, are there days when you must, or must not, eat whale meat? 4. From where and in what way is whale meat transported to you? 5. How do you obtain whale meat? 6. How do you prepare and eat whale meat? In Izukawa's own table of the results from communities (districts) that replied that they ate whale meat, responses to questions 4 and 5 have been collapsed into one, making a total of five categories in all.

I need to add a few words of explanation before using the results of this survey as research data. The first concerns the selection of locations to which the survey was sent. In regard to this point, Izukawa comments, 'it was very convenient for the mailing out of the surveys that, for the most part, I was able to use the local primary schools that had cooperated in our fish trap survey (Izukawa 1942: 115 (1973a: 411)). The 'fish trap' in question seems to be a reference to the *ue*, an implement woven out of bamboo for catching fish, that had been the object of an earlier survey by the Attic

Museum, established by Shibusawa Keizō, and to which Izukawa was affiliated (Yamaguchi 1973). The fish trap project of 1937 was part of the Attic Museum's research on articles for everyday use, following on from its initial research into *ashinaka zōri* (woven straw sandals with no sole under the heel), carried out in 1935 (Miyamoto 1972). The fish trap research not only involved collecting samples of the trap in the field, but also incorporated a questionnaire that 'requested the cooperation of like-minded volunteers from all over the country' (Miyamoto 1972: 964). Apparently, 604 responses to the survey had been received by the end of 1938.

However, the fish trap research was discontinued when several of the researchers who had been involved in the project left to join the Museum of Ethnology and the Institute of Ethnology that opened in 1937 as an adjunct to the Japanese Society of Ethnology (Miyamoto 1972: 964, 967). Consequently, the fish trap research was never published, so we have no way of knowing what sort of questionnaire it was. It is possible that the list of addresses to which the questionnaire was sent still exists somewhere in the archives of the Attic Museum, but up to this stage I have not been able to find out any more about it.

Given these circumstances, the best we can do is to extrapolate from reports of similar surveys made at around the same time. For example, from June to July 1938, Miyamoto Keitarō, who had participated in the fish trap project, conducted research on the types of hats worn in the Japanese home islands, employing a questionnaire asking respondents to answer questions printed on a return-postage-paid postcard. The questionnaires were sent to the headmasters of local primary schools for distribution. At the time, 1503 questionnaires were distributed (responses were returned for 378) to locations Miyamoto selected with the aid of a map according to certain geographical factors. In those areas he considered typical he would select a minimum of two locations per county (*gun*). Then, after the survey had been completed, if the response rate for a particular locality was extremely low, he would seek assistance from his acquaintances in the area to conduct another survey (Miyamoto 1940: 315–46). From this account, we can guess that the selection of locations to be surveyed for the fish trap research was also carried out in the same fashion as that for Miyamoto's own investigation of hats. That is to say, that rather than being a random sample, there was an element of arbitrary decision making involved in the choice of locations. So, these questionnaire surveys of Izukawa and the group affiliated with the Attic Museum, despite their best intentions, do not stand up to rigorous statistical analysis.

Nevertheless, since, as I mentioned before, this is the only pre-World War II survey of the extent of whale meat consumption in Japan, it makes some sense for us to analyse the data from the quantified results of the questionnaire.

In my actual analysis, when considering the responses to the questions, I take each one singly and treat them only in proportion to the number of responses as a whole. I do not attempt to analyse the interrelatedness of more than one question at a time. Even so, I think this method reveals some trends that would not be apparent simply from listing the individual responses.

In addition to his table of responses, shown above, Izukawa includes a map marked with the locations (city, town or village name) surveyed. There are several discrepancies between these two and, moreover, in some places it is necessary to massage the numbers a little in the process of analysing the results. I have indicated these points in the following text and the notes to the table I have put together based on the results of Izukawa's survey.[7]

It is also unfortunate that this research is limited to the Kinki and Chūbu areas. The Kantō area, where, as I mentioned in section two of this chapter, whale meat was not so popular, those areas in northern Honshū and Hokkaidō where whaling stations had been established from early in the twentieth century and Japan's then colonial territories are not included.[8] Nevertheless, the fact that the landlocked prefectures, with the exceptions of Gunma, Saitama and Tochigi, are included in the survey, will, I believe, help us to get some idea of how whale meat came to be eaten in those areas that had not made use of whales in the past, because they had had no access to beached and stranded whales, or because the means of preservation and distribution were insufficiently developed, or for some other reason. Incidentally, in regard to Izukawa's division of whale meat into the two categories of 'blubber' (*aburaniku*) and *akaniku* (red meat), I think it is safe to regard 'blubber' as being the same as the *shironiku* (white meat) I described at the beginning of this chapter.

On the proportion of whale-eating areas

The names of the locations that returned responses to the questionnaire are shown in the table and on the map (for an explanation of the few points of difference and the adjustments to the figures

Table 4.2: Proportion of localities that eat whale meat

	Eat	Don't eat	Eat %
Kyoto	15 + 2	5	75 (77)
Osaka	14 + 1	1	93 (94)
Wakayama	12 + 8	4	75 (83)
Mie	19 + 10	2	90 (94)
Nara	6 + 2	3	67 (73)
Shiga	4 + 2	7	36 (46)
Fukui	7 + 3	2	78 (83)
Gifu	28 + 5	9	76 (79)
Aichi	4 + 2	9	31 (40)
Shizuoka	5 + 2	5	50 (58)
Nagano	31 + 5	2	94 (95)
Yamanashi	13 + 2	4	76 (79)
Ishikawa	10 + 3	2	83 (87)
Toyama	14 + 4	8	64 (69)
Niigata	24 + 14	0	100 (100)
Totals	206 + 65	63	77 (81)

Notes:

1. The data for the survey were collected by filling in the responses for each locality into the table and map, but there are various inconsistencies between the table and the map. Specifically, (1) 'localities where there was a response that whale meat was not eaten' were recorded on the map, but, apart from a few exceptional cases, were not recorded in the table. (2) respondents were asked to name where the locality in question obtained its whale meat, and the source of the whale meat was marked on the map as a 'whale meat-eating locality,' even though that locality had not been surveyed (or there was no response). (For example, when certain communities responded that they sourced their whale meat from Nagoya City, that city itself was marked as a 'whale meat-eating locality' in spite of the fact that Nagoya was not surveyed (there was no response). (3) there are some differences between the map and the table in the names of the locality surveyed (the table lists localities by city, town or village) and whether the response was 'eat' or 'don't eat.' Further, there are localities not recorded in the table, but marked on the map as 'whale meat-eating localities.' Taking these discrepancies into account, I used the following method to analyse the questionnaire. (1) First I confirmed from the map the number of 'don't eat' localities. (2) Next, I compared the 'eat' localities on the map with the table. Then for the 'eat' category, I took 'areas which we know from the table to be whale meat eating localities' plus 'areas from which whale meat is obtained,' which are only on the map. Moreover, when the map has the name of the city, while the table has a number of place names within the city (e.g. the map has Kyoto, but the table has a number of place names within Kyoto city), to avoid duplication, in the case of 'localities from which whale meat is obtained' which appear only on the map I omitted the city (i.e. Kyoto in the previous example). (3) The percentage of localities that eat whale meat was calculated using the following two methods, T/R or (T+M)/(R+M), where T= the number of areas designated whale meat-eating area in the Table; M= areas from which whale meat is obtained (shown only on the Map) and R= the number of Responses to the

questionnaire. The latter result is given in brackets and percentages are rounded to the nearest whole number. (4) There are some discrepancies between the number of responses to the questionnaire (R) and the total of 'areas that we know from the table to be whale meat-eating localities' (T) together with 'localities where there was a response that whale meat was not eaten.' In his case, instead of using the method set out in 3 above, the proportion of whale meat-eating communities is calculated by the following methods, $T/(T+N)$ or $(T+M)/(T+N+M)$, where N= number of localities that responded that whale meat was not eaten (T= Table and M=Map as in 3). Moreover, if a respondent has eaten whale meat even once in the past the village of that respondent is labelled a whale meat-eating village.

2. Where there was confusion, which could not be resolved as an omission or a misprint, between the table and the map regarding place names or whether or not whale meat was eaten, I treated it as follows.

Fukui: I assume Goka in the table corresponds to Uchinami on the map. I base my assumption on the fact that Kamiuchinami and Shimouchinami are place names in present-day Ōno city which includes the former Goka village.

Gifu: Based on the fact that there is a Tase in present day Fukuoka town and that at the time of the survey Yōrō included the former Sawada village, I have taken Fukuoka and Yōrō of the table as equivalent to Tase and Sawada on the map.

I believe Kuguno in the table corresponds to Ōmachi on the map. The location on the map matches the description in the table and it is possible that Ōmachi is a misprint for Ōnishi, a place name in present day Kuguno town.

Sakauchi in the table is probably the same as Sakauchi village today, but there is no record on the map at the location of Sakauchi village. As this is likely to be an omission on the map, I have counted Sakauchi among the 'areas that we know from the table to be whale meat-eating localities.'

Aichi: I take Hōraiji and Sono in the table to correspond respectively to Furi and Komadate on the map. The former is based on the fact that Furi is an extant place name in the town of Hōraimachi which included the former village of Hōraijimura. In the latter I conclude that either the entry in the table or on the map is incorrect, since there is a Komadate in Toyone village bordering on present-day Tōeichō which includes the former Sono village.

Nagano: Minami Otari and Toyosato in the table are both former village names in present-day Otari village and Ueda city respectively, but there are no entries on the map corresponding to their locations. As these may be omissions on the map, I have counted them as 'areas that we know from the table to be whale meat eating localities.'

In regard to Taki in the table, there is no city, town or village of that name in former or present-day Nagano prefecture. However, as this may be a misprint, rather than rejecting it, I have counted it in the category 'areas that we know from the table to be whale meat eating localities.'

In the table Mizuho is marked as a whale meat eating village, but on the map it is marked 'don't eat.' In this case, I have been swayed by the fact that the table lists the type of meat and the route of transmission to include this in the 'eat' category.

Yamanashi: Yamashiro of the table is presumably the then Yamashiro village, but there is nothing recorded on the map at this location. As this may be an omission on the map, I have counted it among 'areas that we know from the table to be whale meat eating localities.'

Ishikawa: Hōryū and Minamiōmi of the table I take to correspond with Ukai and Hachino of the map. Ukai is a place name in present-day Suzu city which includes the former Hōryū village and Hachino is a place in the present-day township of Takamatsu which includes the former village of Minamiōmi.

3. In the Osaka section of the table mention is made of Kōbe, and Amagasaki is marked within the area of Osaka prefecture on the map, but these have been disregarded.

required to correct these problems see the notes to Table 4.2). Let us now see from this data just what proportion of the locations surveyed ate whale meat. According to the table 77 [81]% of all the localities surveyed ate whale meat. But, we can see a degree of dispersion in the figures from one prefecture to another. At the low extreme we have Aichi and Shiga prefectures with 31 [40]% and 36 [46]% respectively. These are followed in order by Shizuoka, Toyama and Nara. We can say that these five prefectures fall within the lower range compared with the other prefectures. I think we can surmise that the reasons behind the extremely low figures for Aichi and Shiga are as follows. First, as Aichi had a flourishing poultry farming industry,[9] we might think that chicken meat had established a strong foothold throughout the prefecture, leaving little room for whale meat to penetrate the market. But it is debatable whether the consumption of chicken meat, which was still regarded as a luxury food item before the Second World War (See Note 4), would have been high enough (Yoshioka, Ōnishi Nobuhiko *et al.* 1963: 16–17, 194–6) to account for the low consumption of whale meat. Next, to take the case of Shiga prefecture, the low consumption of whale meat may be attributed to the fact that although Shiga has no seacoast, it has a massive lake, Lake Biwa, within its borders. Consequently, the locals make good use of the abundance of fish and other freshwater products from the lake. We can assume that in these circumstances the people did not develop a taste for whale meat and saw little reason to choose it in preference to what was readily available. This would explain the extremely low rate of whale meat consumption in Shiga prefecture. Moreover, to judge from Izukawa's map, the communities in Aichi prefecture that returned completed questionnaires are concentrated in the Mikawa area, with no responses for Owari. It is possible that the results may have been a little different if Owari had been included. There is an interesting research report regarding the reason for the comparatively low rate of whale meat consumption in Shizuoka, but I deal with this later.

Now, if we turn out attention to the high end of the spectrum, we find Niigata with a figure of 100%. It is interesting to note that the next highest, with a consumption rate of 94 [95]%, is the land-locked prefecture, Nagano. The table and the map both indicate that whale meat was brought into Nagano via Niigata prefecture, Tokyo, Nagoya and so on. Further, mention should be made of the surprising fact that four communities in Wakayama,[10] where net-whaling had been practised in the past and whaling companies later engaged in Norwegian-style whaling, claimed they did not eat whale meat.

When was whale meat introduced?

From this point on, I examine the actual responses to the questions in the survey, concentrating on those communities with an entry in Izukawa's table. My classification of the period of introduction of whale meat into each of the communities is summarised in Table 4.3. First, I need to explain the basis for my chronological divisions in the table. In Chapter 1, I divided the development of the Japanese whaling industry into five periods, bounded by these events; the introduction of Norwegian-style whaling in 1897, the birth of Tōyō Whaling (Tōyō Hogei) in 1909, the first factory-ship whaling of 1934 and the discontinuation of factory-ship whaling from 1942. Of course, this last date can be ignored as it is later than Izukawa's 1941 survey. So, working back from 1941, the year of the questionnaire, these should equate with the responses 44 years ago, 32 years ago and 7 years ago. But inspection of the actual answers given in the questionnaire shows that there are respondents in particular who answered, 'about eighty years ago,' or 'fifty years ago,' or 'thirty years ago.' There were also some that said 'about ten years ago.' When relating the period given in the responses to what happened at the time, for example, we can think of 'about fifty years ago' or 'around ten years ago' as referring respectively to when whale meat, obtained through the advent of Norwegian-style whaling in the former case, or in factory-ship whaling in the latter,[11] first came to be consumed in that particular community. So, rather than mechanically following my historical chronology and to ensure that answers like 'fifty years ago' and 'ten years ago' are not disregarded, I equate the beginning of the period with a response of 'fifty years ago' and the introduction of factory-ship whaling with a response of 'ten years ago.' For thirty-two years ago, I have substituted thirty years ago (which in any case would be a statistically valid response for thirty-two years), to preserve the balance of ten-year divisions.

Moreover, the numerous replies along the lines of, 'eighty years ago,' or 'from the beginning of the Meiji period,' can presumably be explained as a reference to the relaxation of the taboo against eating meat, that accompanied the social changes following the Meiji Restoration of 1868. To include these replies in the analysis I have made a division at eighty years and the final category, more than eighty years ago.

Even so, a little over one third of the responses to this question are 'unclear' or 'don't know.' So, I continue with my analysis, mindful of Izukawa's own conclusion that 'there is some doubt

Table 4.3: 'When did whale meat first enter your community?' (counting back from 1941)

	Kyoto	Osaka	Wakayama	Mie	Nara	Shiga	Fukui	Gifu	Aichi	Shizuoka	Nagano	Yamanashi	Ishikawa	Toyama	Niigata	Totals
In last 10 years	1 (7)	0	0	0	0	0	0	0	2 (50)	1 (20)	0	0	0	0	0	4 (2)
11 to 30 years ago	0	2 (14)	1 (8)	4 (21)	3 (50)	2 (50)	0	8 (29)	2 (50)	0	3 (10)	5 (38)	0	6 (43)	2 (8)	38 (18)
31 to 50 years ago	4 (27)	5 (36)	1 (8)	3 (16)	1 (17)	0	1 (14)	3 (11)	0	0	7 (23)	2 (15)	0	1 (7)	3 (13)	31 (15)
51 to 80 years ago	0	2 (14)	2 (17)	2 (11)	0	0	0	1 (4)	0	1 (20)	3 (10)	2 (15)	2 (20)	1 (7)	3 (13)	19 (9)
More than 80 years ago	7 (47)	1 (7)	4 (33)	1 (5)	1 (17)	0	2 (29)	4 (14)	0	0	3 (10)	1 (8)	1 (10)	4 (29)	12 (50)	41 (20)
Don't know	3 (20)	4 (29)	4 (33)	9 (47)	1 (17)	2 (50)	4 (57)	12 (43)	0	3 (60)	15 (48)	3 (23)	7 (70)	2 (14)	4 (17)	73 (35)
Total	15	14	12	19	6	4	7	28	4	5	31	13	10	14	24	206

Notes:

1. The figures in parentheses ( ) are percentages, ie. the numbers for each category divided by the total for the prefecture, or, in the case of the final column, the total of locations.

2. This table represents the responses to the question, 'when did whale meat enter your community?' indicating when whale meat was first eaten, regardless of the quantity

or type of meat, but probably excludes the use of certain internal organs of the whale taken for medicinal purposes.

3. Vague expressions like 'about such-and-such a year', 'around so many years ago' or 'we were eating it so many years ago, but I don't know how long before that' were analysed as if the consumption of whale meat had started in the year mentioned.

4. The 'don't know' category includes expressions like, 'from long ago,' 'for a very long time,' 'a ninety year old man said he used to eat it when he was a child.'

5. The responses, 'from the end of the Meiji period,' 'in the latter half of the Meiji period' and 'after the Sino-Japanese War' were included in the '31 to 50 years ago' category. 'Since Meiji,' 'in the early years of the Meiji period' etc came under '51 to 80 years ago' and 'before Meiji' 'before the beginning of the Meiji period,' were grouped into the 'more than 80 years ago' category.

6. The expression 'fifty or sixty years ago' has been included in the 51 to 80 years ago category.

over how much trust we can place in this report' (Izukawa 1942: 116 (1973a: 412)).

First, seen as a whole, we find that a total of 73 communities (35%) responded that the eating of whale meat began after the introduction of Norwegian-style whaling, i.e. within the last fifty years. This figure combined with those in Table 4.2 that 'don't eat' whale meat comes to 136 communities. This is 49% (all percentages are rounded to the nearest whole number) of the total number of responses (279) in Table 4.1 and 51% of the total of 'eat' communities (that are listed in Izukawa's table only) and 'don't eat' communities (269) that are recorded in Table 4.2. That is to say, at least half of the communities that returned completed questionnaires did not eat whale meat before the introduction of Norwegian-style whaling.

Next, let us consider the figures from each prefecture separately. To begin with those five prefectures, introduced above, which had a relatively low rate of whale meat consumption, and also bearing in mind that both Shiga and Shizuoka had a high proportion of 'don't know' responses, we notice the trend in these five prefectures overall indicating that the eating of whale meat began comparatively recently, with the majority of communities returning the response, 'within the last thirty years.' The 'land-locked' prefectures also have high percentages of communities claiming whale meat came in no more than thirty years ago. In Nara and Shiga 50% of communities gave responses indicating entry within the last thirty years. The figure for Yamanashi was 38% and for Gifu 29%, both also comparatively high. However, Nagano, with a high rate of whale meat consumption, returned a low 10% of 'within the last thirty years' responses, and even 'within the last fifty years' accounted for only 32% of the responses, just a little under the proportion for the data overall (35%, mentioned above). But Nagano has a high proportion of 'don't know' responses (48%) and the 19% figure for responses placing the arrival of whale meat before the introduction of Norwegian-style whaling, i.e. more than fifty years previously, is below the average for all prefectures (29%). Consequently, judging from these figures, I think we can safely say that whale meat came into the land-locked prefectures in a comparatively recent period.

Now let us look at the other prefectures. In Niigata, where 100% of communities said they ate whale meat, the introduction of whale meat was early, with 50% of the communities surveyed responding that whale meat had come in more than eighty years ago. In Wakayama, too, the figure of 33% for responses indicating more than eighty years since the introduction of whale meat was also rather on the high side. Kyoto with 47% in this same category was also high. But according to the map, four of the communities surveyed in Kyoto faced the Japan Sea. It is known that in Ine in the Tango district of present-day Kyoto, from around 1600 until the end of the nineteenth century, whales were taken with the fishing technique known as *tatekiriami* (literally, shield-cutting net), in which the bay was sealed off with nets (Fukumoto 1978: 41–3, 52–4). In Izukawa's composite table, there is a response to the effect that in Ine and Kunda, at this time they made use of whales they caught locally (fishing method unknown).[12] On the other hand, 27% of respondents answered 'between 31 and 50 years ago.' This figure, taken together with the 36% for this category in Osaka, can be interpreted as lending support to my earlier observation that whale red meat (*akaniku*) seems not to have been widely eaten in the Kansai (Ōsaka–Kōbe) area prior to the advent of Norwegian-style whaling.[13]

The type of whale meat eaten

Table 4.4 shows the type of whale meat eaten, adding the category of canned whale meat, which turned up frequently in the responses, to the *akaniku* (red meat) and *aburaniku* (blubber) of the questionnaire. Taken overall, the most frequent response, making up 31% of the total, was that both types of meat were eaten. This was followed by red meat only (25%), blubber only (23%), and only canned whale meat accounted for 10% of the total. Moreover, the total of 125 locations that reported the eating of red meat make up 61% of the 206 locations in Izukawa's table that reported that they ate whale meat, 45% of the 279 responses in Table 4.1 and 46% of the 269 combined responses 'eat' and 'don't eat' in Table 4.2. That is to say, red meat was consumed in only around 45% overall of the locations surveyed and in around 60% of those communities that ate whale meat.

Next, considering those prefectures that returned a low percentage of 'eat' responses, we cannot identify any particular trend of shared characteristics. We can say, however, that the eating of canned whale meat is concentrated into the land-locked prefectures. In particular, the figures reveal a trend, in those land-locked prefectures, for locations that reported that they only ate canned whale meat to become more numerous as we move further east. Taking into account the circumstances of canned food production explained earlier in this chapter, we can say that this adds support to the findings of our analysis of Table 4.3, that whale meat was introduced comparatively recently into most of the land-locked prefectures. I think we can also claim that the high percentage of 'canned whale meat only' in Nagano and Yamanashi prefectures indicates that whale meat was not yet eaten on a regular daily basis.

In addition, it is interesting to note that in comparison with the percentages overall, the proportion of 'red meat only' in those prefectures on the Pacific coast like Osaka, Wakayama and Mie (also Aichi and Shizuoka) and 'blubber only' in Fukui and Niigata on the Japan Sea coast are extremely high. Later, I would like to make some observations regarding possible reasons for this.

Days when whale meat is, or is not, eaten

Finally, I would like to analyse what we might call 'folklore factors,' looking at the responses to the question, 'Are there days when you must, or must not, eat whale meat?' First, in regard to the 'dog days of summer' (*doyō*), as we can see from Table 4.5, while there were no communities on the Pacific coast that claimed to eat whale meat at this time, there were communities in Kyoto, Fukui, Ishikawa and Niigata facing the Japan Sea and in the land-locked prefectures, Gifu, Nagano and Yamanashi, that said that they did. The percentages in Fukui and Niigata were particularly high. These findings overlap with those of Table 4.4, which showed that the proportions of 'red meat only' on the Pacific side and 'blubber only' in Fukui and Niigata on the Japan Sea side were both extremely high. Actually, according to a 1912 report, demand for *shironiku* ('white meat') had spread across the country and was particularly relished as *doyō kujira*, whale meat to be eaten in the 'dog days' at the height of summer (Andō 1912b: 29–30). In Fukui prefecture, where almost all of the communities responded that whale meat was eaten at *doyō*, and also in the descriptions for Fukui in Izukawa's table, apart from one community that ate red meat only, all said they ate whale meat at *doyō*, served in miso soup or some other kind of broth. In Niigata, too, of the ten communities that replied that they eat whale meat at *doyō*, three said they ate both red meat and blubber, one indicated blubber or canned whale meat and the remaining six all responded 'blubber only.' In regard to mode of preparation, all ten communities responded that they used the whale meat in soup.

Now, based on these facts, I think we can conclude as follows. In many places along the Japan Sea coast, whale white meat was eaten as a 'lucky food' (*engimono*) for good health, in much the way as eel is, and for some reason this custom also spread to the land-locked prefectures of Gifu and Nagano. Conversely, I think we can also say that those areas that ate 'white meat only' certainly did not eat whale meat as part of their daily routine. Incidentally, although they both face the Japan Sea, the number of communities in Kyoto and Ishikawa that indicated they ate whale meat at *doyō* was rather low and there were no reports at all of this custom from Toyama. What these three prefectures have in common, as we can see from Table 4.4 is that there is a high proportion of communities that responded that they ate both red meat and blubber and that there were also communities that said they used whales caught locally (See Note 12 of this chapter). As the designated whaling season around the Noto Peninsula was from winter through to the end of spring (Kitamura 1838: (1995: 132–4); Tōyō Hogei K.K. 1910: 20), we can assume that the same was true for the Japan Sea coast as a whole. This leads us to think that possibly communities in these three prefectures consumed whale meat (red meat) during the winter months on a relatively more regular, daily basis than other prefectures on the Japan Sea. We can speculate, therefore, that this was the reason that it was not regarded here as a special lucky food. We can probably say the same for the Pacific coast, particularly for those communities that engaged in whaling.

Incidentally, if we apply the same calculation used in our analysis of red meat in Table 4.4 to blubber we come up with a total of 124 communities, or roughly the same number as with red meat. From this we see that forty per cent of the locations in Izukawa's table that responded that they ate whale meat made use of white meat. We can divide the remainder, i.e. those who don't eat white meat, into two groups, those who eat only red meat and those who eat only canned whale meat. In the prefectures with a relatively low rate of whale meat consumption, and in the land-locked prefectures, we can assume that the responses, 'red meat only' or 'canned whale meat only' indicate a comparatively recent date of introduction – that is to say, dating from the time when whale meat was introduced into areas where it had not been previously eaten, as a result of active promotion by the powerful, whaling industry and nationalist policies of the State.[14] It is not clear why 'red meat only' should be relatively high in Wakayama and Mie prefectures where whaling was practised. Could it perhaps have been because red meat was favoured in communities close to whaling areas and white meat, which was preserved in salt, was kept for sale to distant areas?

Table 4.4: Type of whale meat eaten

	Kyoto	Osaka	Wakayama	Mie	Nara	Shiga	Fukui	Gifu	Aichi	Shizuoka	Nagano	Yamanashi	Ishikawa	Toyama	Niigata	Totals
Only red meat (akaniku)	1 (7)	6 (43)	7 (58)	12 (63)	1 (17)	1 (25)	1 (14)	8 (29)	3 (75)	3 (60)	2 (6)	1 (8)	1 (10)	4 (29)	1 (4)	52 (25)
Only blubber	2 (13)	0	0	1 (5)	1 (17)	1 (25)	5 (71)	10 (36)	0	0	10 (32)	0	3 (30)	1 (7)	13 (54)	47 (23)
Only canned whale meat	0	0	0	0	0	0	0	3 (11)	1 (25)	0	9 (29)	7 (54)	0	1 (7)	0	21 (10)
Both red meat and blubber	10 (67)	8 (57)	4 (33)	6 (32)	2 (33)	2 (50)	1 (14)	3 (11)	0	2 (40)	5 (16)	2 (15)	6 (60)	7 (50)	5 (21)	63 (31)
But mainly red meat	1 (7)	1 (7)	2	1	0	0	0	1	0	0	0	0	1	0	1	8
But mainly blubber	0	0	0	1	0	2	1	1	0	0	2	1	0	0	2	10
Red meat and canned whale meat	0	0	0	0	0	0	0	1 (4)	0	0	1 (3)	1 (8)	0	1 (7)	0	4 (2)
Blubber and canned whale meat	0	0	0	0	0	0	0	1 (4)	0	0	3 (10)	0	0	0	4 (17)	8 (4)
Red meat, blubber and canned whale meat	0	0	0	0	1 (17)	0	0	1 (4)	0	0	1 (3)	2 (15)	0	0	1 (4)	6 (3)

																Total
Don't know	2 (13)	0	1 (8)	0	1 (17)	0	1 (4)	0	0	0	0	0	0	0	0	5 (2)
Total	15	14	12	19	6	4	28	7	4	5	31	13	10	14	24	206

Notes:

1. The figures in parentheses ( ) are percentages, ie. the numbers for each category divided by the total for the prefecture, or, in the case of the final column, the total of locations (rounded to the nearest whole number). Consequently, totals do not always add up to exactly 100%.

2. The categories include instances where a respondent has eaten a particular type of whale meat only once in the past. The eating of whale organs has not been set up as a separate category as it rarely came out in the responses and was for medicinal purposes. These responses were ignored.

3. Responses along the lines of 'mainly red meat' or 'more often blubber' were included in the category 'both red meat and blubber,' then further categorised individually.

4. Responses regarding the nature of the whale meat eaten, given in reply to other questions, such as the way the meat was cooked or the period when whale meat was introduced into the community, were also included the above categories.

5. When 'canned whale meat' was given as the only response to the question on cooking methods it was included in the category 'only canned whale meat,' regardless of the answer to the question on types of whale meat eaten, unless it was obvious from answers to other questions in the survey that whale meat other than the canned variety was eaten.

6. The responses like 'koro', 'irigara' and 'ijigara' apparently refer to 'whale oil dregs' (*irikasu*), which occur as a by-product of oil extraction. Consequently, I have included them under 'blubber.'

7. The responses 'skin' (*kawa*) and 'kuro' (literally, 'black') have been included under 'blubber.'

8. I have taken the response 'raw whale meat' to indicate red meat (*akaniku*).

9. Similarly, 'dried whale meat' is regarded as red meat.

10. 'Oily red meat' (*abura akaniku*) I have assigned to the category 'both red meat and blubber.'

11. The two communities in Nara and Niigata prefectures (Hatano and Teradomari respectively) that answered that they eat red meat, blubber and canned meat, mainly eat blubber. The community in Gifu that also claimed to eat all three types (Okumyōgata), it seems, relies predominately on canned whale meat.

Table 4.5: Days on which whale meat is, or is not, eaten

	Kyoto	Osaka	Wakayama	Mie	Nara	Shiga	Fukui	Gifu	Aichi	Shizuoka	Nagano	Yamanashi	Ishikawa	Toyama	Niigata	Totals
Whale meat eating communities	15	14	12	19	6	4	7	28	4	5	31	13	10	14	24	206
Communities that eat it at 'dog days' of summer	3 (20)	0	0	0	0	0	6 (86)	5 (18)	0	0	7 (23)	1 (8)	1 (10)	0	10 (42)	33 (16)
Communities that eat it on festival days	0	1	0	0	0	1	0	0	0	0	0	0	0	0	0	2
Communities that eat it at weddings	0	0	0	0	0	0	0	0	0	0	0	0	0	0	3	3
Communities that don't eat it on days of Buddhist rituals	1 (7)	1 (7)	3 (25)	1 (5)	0	0	0	0	0	0	0	0	1 (10)	1 (7)	1 (13)	11 (5)

Notes:

1. Percentages in parentheses ( ) are rounded to the nearest number.

2. The categories include instances where a particular custom prevailed in the past.

3. Vague responses like, 'there are some families where they eat it on the day of the ox during *doyō*' (dog days) are interpreted as a community that eats whale meat during the 'dog days' of summer.

4. There are places where the response to the question, 'Are there days when you do, or do not, eat whale meat' is recorded as, 'on the day of the ox during *doyō*' or similar, without specifying whether whale meat was eaten or not. I have taken these to mean that whale meat was eaten, as there is no record clearly stating that whale meat is not eaten during *doyō*.

5. I have judged that responses to this question along the lines of, 'we mainly eat it in summer' or 'we eat it in July,' or 'we eat it to beat the fatigue of the summer heat' or 'it helps you cope with the heat' as belonging to the eaten during *doyō* category.

6. There is one response for Kōbe, which we have rejected because it is mistakenly included under Osaka in the table, which states that 'in the past it was apparently used in a collation mixed with vegetables (*aemon*) served on festival days.'

7. One response from Niigata (Kamigō) states, 'Under the old lunar calendar it is always eaten on the 31st December, and the 15th January and is never eaten on 11th January, 2nd May, 4th September or the 4th December.' But apart from New Year's eve, it is not clear what these days signify, so I have chosen not to include them under festival days or days of Buddhist ceremonial.

8. I have included the responses, *shūgen* (marriage), *yometori* (taking a bride) and *mukotori* (taking a groom) under 'wedding ceremony' (*kekkonshiki*).

9. I have included under 'days of Buddhist ritual' the responses, '*hotokebi*' (Buddha days), '*shōjinbi*' (days of abstinence), 'anniversary of the death of a parent or family member,' '*kōshin no hi*' (the day of the monkey in the elder-brother-metal cycle of the traditional calendar) and '*imibi*' (taboo days).

Two locations reported that whale meat was eaten on festival days *(matsuri no hi)*. One of these was Fuse in Osaka. There is a note stating, 'After completing the ceremony before Oyumi Sama (the sacred bow) on the ninth day of January, there is a banquet in the house of the festival manager and whale meat is always included in the foods eaten at this time (along with *mizuna* (potherb mustard)), but the reason for this custom is not known.' The other is Mikami in Shiga Prefecture, from which we have a response stating that, 'every year on the ninth day of February, on the festival day of the shrine, whale meat is always served as one of the festival dishes *(omono)* and eaten by the teams of men who carry the *mikoshi* (portable shrine). This custom started long ago and is still followed today. Further, it seems that there is one place in the vicinity where whale meat is eaten every year without fail in the middle of January.' Further, the only responses claiming that whale meat was eaten at weddings came from three locations in Niigata, all claiming that it was used in soup. According to other sources, whale meat is invariably among the foods included in wedding banquets and other auspicious occasions in the Hyūga district of south-east Kyūshū. And in Nagasaki city, 'every house has to make sure it has a supply of whale meat to have as a side dish with *sake*' at new year, *setsubun* (the day before the beginning of spring), and, in agricultural areas, at the time of rice planting (Tōyō Hogei K.K. 1910: 136–7). There is also a report that, in Kyūshū and in north-eastern Honshū, at the time of the new year celebrations, as a symbol of good luck, a piece of whale skin is placed on the tray alongside the rice cakes for the *zōni* soup.[15] The reason suggested for this is that perhaps it is considered lucky to eat something big at New Year; so salmon is preferred over sardines and whale over yellowtail *(buri – Seriola quinqueradiata)* (Andō 1912b: 30). This was probably also the reason that it was considered auspicious as a food for festivals or wedding banquets.

In contrast to these, there were eleven locations that reported that whale meat 'must not be eaten' on days of so-called Buddhist rituals. It was not possible, however, to recognise any particular trend in the distribution. Among the responses, for example, were some along the lines of, 'as with fish, whale is not eaten on Buddhist days' (Yuasa in Wakayama) or 'meat is not eaten on taboo days' (Yunotani in Niigata). Moreover, there were also clearer responses like, 'in the past it (whale) was not eaten as its powerful effect on the mind made it unsuitable for Buddhist worship. It was not eaten until the Meiji period' or 'nowadays we no longer make a distinction between days (when it may and may not be eaten)' (both of the former responses are from Shichikawa in Wakayama). Rather than indicating customs related to whales in particular, these responses should probably be interpreted as indicating that, at this time, the custom of avoiding meat when taking part in Buddhist ceremonies was still prevalent across the whole country.

Here, in regard to this point, I venture to take up some comments appended to Izukawa's table in connection with the 'don't eat' responses. The comments in question concern two responses from Shizuoka prefecture. There is a report from Omaezaki to the effect that 'whaling, let alone the eating of whale meat, is alien to this area. I have not even heard any tales or legends about whales here' and another from Shimoda stating that, 'in the Izu district, from long ago there has been a belief that whales are gods of good fortune and so they are never hunted and there is no custom of eating whale meat, though dolphins are eaten.' We can regard this report as being another example of an area with the belief, introduced in Chapters 1 and 2, that whales are the god Ebisu. But in this case it is necessary to make a minor qualification. First, in regard to the reason whales were regarded as gods of good fortune, we have already seen that the origin of the Ebisu belief stems from the fact that whales chase the sardines the fishermen are seeking to catch and drive them closer to shore. I think we can find a similar reason for the belief in Izu. The area along the coast of Shizuoka prefecture from Suruga Bay to the Izu Peninsula is renowned for the rod and line fishing of bonito. It is reported that, in order to locate schools of bonito, fishermen would strive to see not only flocks of streaked shearwaters or floating logs, which are possible indications of shoals of bonito, but also (sei) whales (Shizuokaken 1989: 73, 125–6; 1991: 477–83).

In this case, however, we need to be a little more cautious. Moving from west to east along the stretch of coast from Suruga Bay to the Izu Peninsula, we have responses from the five locations of Omaezaki, Shimizu-Komagoe (Izukawa's map has Seimizu, but this must be a misprint for Shimizu), Nagahama, Shimoda and Ajiro. Of these, the situation in Omaezaki and Shimoda was as I have described above. But there are reports from Shimizu-Komagoe to the effect that they eat whale meat brought in from outside and from Nagahama and Ajiro that they eat the meat of whales they catch in the vicinity. We can surmise that the reason for this discrepancy might have been confusion from one locality to another or one individual to another as to whether or not to include dolphins in the same category as whales, for, as we saw in the report from Shimoda, there is a long history of dolphin fishing in the Izu Peninsula (See, for example, Nakamura 1988: 100–23).[16] So it is very possible that in the judgement of some of the respondents to the questionnaire, dolphins were seen as whales. Further, we could guess that perhaps at around this time, the taboo on eating whale meat was also gradually being relaxed. It is difficult to comment further on this point,[17] but, in any case, in this area, I think we can at least say that it is possible that, of the eighty species in the order cetacea said to exist today, including dolphins and porpoises, some were treated differently from others.

Summary

While conceding that the questionnaire had its limitations, I believe it has helped us clarify a number of issues in the debate so far. I'll recap on these points directly. The stimulus for the spread of whale meat consumption in modern Japan was the introduction of Norwegian-style whaling. From then on, under the aggressive promotion of the whaling industry, which had soon consolidated into a huge monopolistic enterprise, together with the strategic activities of the State for the purpose of waging war, the eating of whale meat, which even in the Kinki and Chūbu areas of western and central Honshū had previously been eaten in only around half of the communities surveyed, gradually spread throughout the country.

As a result of these factors, by 1941 around eighty per cent of communities in the Kinki and Chūbu areas ate whale meat. When we look closely at the actual figures, however, we discover that in some prefectures there were communities where the consumption of whale meat was extremely low. Further, the analysis showed that the spread of red meat (*akaniku*) extended to only about sixty per cent of those communities that ate whale meat and clarified the fact that the white meat, which had been eaten from earlier times, was often regarded as a lucky food to be consumed only on special occasions. In addition, the statistics showed a strong tendency towards white meat consumption on the Japan Sea coast, red meat on the Pacific coast and canned whale meat in the land-locked prefectures. Moreover, there were some communities that did not engage in whaling or eat whale meat because they regarded whales as gods of good fortune. We can say, then, that the survey revealed that there was a good deal of variation from one community to another in when and why whale meat was eaten, if indeed, it was eaten at all.

On the basis of this evidence, I think we can agree with Morita's observation that it was not until after World War II that whale meat came to be eaten generally across the whole country as a part of the daily diet. Moreover, the survey confirms me in my failure to be convinced by the discourse that holds that the eating of whale meat is part of the traditional food culture of *the Japanese race*.

Further, taken together with the facts presented in Chapters 1 and 2, this chapter clearly shows how the hunting and eating of whales led to a process in which, under the control of the Japanese State, or a State-protected industry, a food item had penetrated areas, including Japan's colonies, where it had not previously been part of the normal diet. In this connection, R. L. Barsh, to whom I referred at the beginning of this chapter, points out how the diets of indigenous societies, in which the eating of products obtained through the hunting of wildlife played an important role in the maintenance of health, were destroyed by the penetration of European diet under Western hegemony (Barsh 2001: 158–65). That was certainly so in the case of indigenous societies, but as we have seen in this chapter, in so much as the penetration of the practices of hunting and eating whales was abetted through the hegemony of the Japanese State, Barsch's schema of hegemony putting an end to the hunting of wild animals, is perhaps a little too simplistic.

5 In Search of the Logic of Overfishing: An Analysis of Whaling Discourse

Introduction

There seems to be no disagreement, even from those involved in the industry (See, for example Nagasaki 1990a: 19–20) that modern whaling, particularly since the Second World War, has indulged in overfishing. How, you may wonder, did it reach this stage? In Chapter 3, I explained the logic and the system of wildlife protection in the modern period. In this chapter, I propose to follow on from there, to investigate the thinking of those actually engaged in whaling, or, to put it another way, to probe the *logic of overfishing.*

First, in the discussion that follows, I take up, for the first time in the present volume, the question of the development of the Japanese whaling industry from the end of World War II to the abolition of the BWU system of whaling in the Antarctic Ocean in 1972. From there, I go on to home in on the *logic of overfishing* in the discourse of those connected with whaling in Japan, centring on the early modern and contemporary periods.

Before I do so, however, I would like to add a few words of explanation about discourse analysis. My understanding of the method as it is set out in Michel Foucault's *L'Archéologie du Savoir*, runs roughly as follows. In short, a discourse is a group of statements, formed from among those statements made in the sense of representing or narrating a certain object.

And the analysis of that discourse is the definition of the system that governs the object, and the rules that form the statements. This technique, which Foucault called 'the archaeology of knowledge,' does not attempt, as the so-called history of ideas does, to place the narrative concerning an object into a unified, coherent framework. Nor, to take an analogy, does it depict it as the flow of a single river. Instead, without attempting to reduce the diversity of the discourse or resolve its contradictions, it sees a discourse as a spatial arrangement, with its differences left unresolved, and it strives to elucidate the factors that dictate that arrangement.

These days, however, the term 'discourse analysis' encompasses not only the technique elaborated by Foucault, but also a range of variations, including so-called textual analysis, conversation analysis in ethno-methodology and press content analysis in the mass media. In these various discourse analyses, Akagawa Manabu emphasises the importance of the social position of the narrating subject, even above what is said about the object of the discourse itself. He points out the tendency to bring into question the latent self-interest of the narrating subject and the political effect of the discourse (Akagawa 2001: 65, 77). And what I call discourse analysis in this chapter (or in the whole of this book, for that matter) follows Akagawa's suggestion, in that it takes up the problem of the so-called politics of representation of the taking of whales and objectification of whales *per se*. In this sense it does not strictly adhere to Foucault's methodology.

Nevertheless, in this chapter, by identifying and analysing a number of statements about the catching of whales made over a long period, I attempt to elucidate one or more general dominant rules underlying the formation of those statements. And in doing so, I also analyse the *rarity* of those statements; that is, I weigh up the poverty of statements and seek out the rule underlying it. In other words, I analyse why it is that certain statements are made and others are not (Akagawa 2001: 77–81; Foucault 1969 (Japanese translation [1995]: 181–5)).

The development of Japanese whaling after WWII

Let me begin with an overview of Japanese whaling from the end of WWII until 1972, based on appropriate reference materials, particularly Hara Takeshi for Antarctic (Hara 1993), and Kondō Isao for coastal, whaling (Kondō 2001). This section should be read in conjunction with Table 5.1.

To take Antarctic whaling first, on 6 August 1946, GHQ (General Headquarters, Supreme Commander for the Allied Powers), gave its formal permission for Japanese whaling fleets to fish in the Antarctic Ocean, overriding the opposition of some of the allies who feared the Japanese navy would revive and the Japanese would engage in indiscriminate harvesting of whales. As a result, two fleets, those of Nippon Suisan Kabushiki Kaisha (Nippon Fisheries) and Taiyō Fisheries (See Table 5.2 for changes in whaling companies in the post-war period), comprising a factory ship which had been hurriedly converted from an oil tanker and the whale catchers that had survived the war (Maeda and Teraoka Yoshirō 1952: 38–9) set sail for the Antarctic. Later, in the 1951–52 season, Nippon Fisheries raised the whaling factory ship *Daisan Tonan Maru* which had been sunk during the war, and after repairing the vessel and renaming it the *Tonan Maru* decided to use it as a factory ship again. Further, Kyokuyō Whaling dispatched a fleet to hunt exclusively for sperm whales (Maeda and Teraoka Yoshirō 1952: 39–42).

Further, in December 1946 fifteen countries signed the International Convention for the Regulation of Whaling and in 1948 the IWC (International Whaling Commission) was established to administer and drive this agreement. When the first meeting of the IWC was held in London in May to June 1949 (Maeda and Teraoka Yoshirō 1952: 251; Shigeta 1963: 15), it was decided to impose an overall quota of sixteen thousand BWU for the Antarctic Ocean. The BWU, or blue whale unit, equates one blue whale with two fin whales, 2.5 humpback whales and six sei whales. Briefly, the way the system worked was that, for the duration of the whaling season, the whales taken by each of the fleets were converted to equivalent BWUs and reported to the Bureau of International Whaling Statistics in Norway. When the total for all whaling fleets had reached 16,000 BWUs this bureau would then order the cessation of whaling.

This method known as the 'whaling Olympics' meant that it was more efficient in terms of the labour expended in catching and flensing to hunt from the larger species down. As a result, the great bulk of whaling pressure fell on the larger species of whale. Moreover, this method applied the same single criterion of quantity across the board regardless of the different ecologies of whale species. Seen from the perspective of one living today (the author, Watanabe), there were problems with this system of whale regulation.

Table 5.1 was compiled by the author (Watanabe) using the following reference material, in addition to the sources covered in the argument so far. A-Team (1992), *Kujira no kyōkun* (The moral of the whale), JMA Management Center; Baba, Komao (1942), *Hogei* (Whaling), Tokyo: Tennensha; Fujishima, Norihito and Yoshiaki Matsuda (1998), 'IWC ni yoru geirui shigen kanri no tayōsei e no taiō ni kansuru ichi kōsatsu' (An observation concerning the response to diversity in the IWC's whale resource management), *Chiiki Gyogyō Kenkyū* (Regional fisheries research), 39 (1), pp. 111–24; Fukumoto, Kazuo (1978), *Nihon hogei shiwa* (Historical tales of Japanese whaling), Hōsei Daigaku Shuppankyoku. reprinted [1993]; Hara, Takeshi (1993), *Za kujira – daigoban* (The whale – 5th edition), Tokyo: Bunshindō; Maeda, Keijirō and Teraoka Yoshirō (1952), *Hogei* (Whaling), Tokyo: Nihon Hogei Kyōkai; Morita, Hideo (1963a), 'Nihon hogeigyō no saihensei wa hisshi ka' (Is a rearrangement of the Japanese whaling industry inevitable?), *Suisankai* (Fisheries world), 941, pp. 30–7; Nihon Honyūrui Gakkai ed. (Mammalian Society of Japan ed.) (1997), *Reddo dēta Nihon honyūrui* (Red data Japanese mammals), Bun'ichi Sōgo Shuppan K. K.; Sakuramoto Kazumi, Katō Hidehiro and Tanaka Shōichi eds (1991), *Geirui shigen no kenkyū to kanri* (Studies on whale stock management), Kōseisha Kōseikaku; Shigeta, Yoshiji (1963), 'Sekai no hogei seido oyobi sono haikei (hachi)' (The world whaling system and its background (8)), Geiken tsūshin (Institute of Cetacean Research newsletter), 137, pp. 12–22; 'Geirui no hokaku nado o meguru naigai no jōsei heisei jūgonen shichigatsu' (Internal and external conditions surrounding the taking of whales etc, July 2003), *http://www.jfa.maff.go.jp/whale/document/ brief_explanation_of_whaling_jp.htm* (accessed on 3 November 2005); 'Kokusai Hogei Iinkai (IWC) tokubetsu kaigō no kekka ni tsuite' (On the results of the special meeting of the IWC), *http:// www.jfa.maff.go.jp/release/14.10.15.4.htm* (accessed on 15 October 2004); Articles of *Geiken tsūshin* (Institute of Cetacean Research newsletter).

Ultimately, there can be no doubt that it was the adoption of this system that led to the drastic reduction in the numbers of large whales in the Antarctic Ocean.

Incidentally, the immediate granting of permission for Japan to resume whaling, mentioned above, was to overcome the food shortage following the war. Figure 5.2 is a graph of the changes in the annual meat production per Japanese citizen (*kokumin*). We can see from this that the proportion of whale meat, which, even in the 1930s, accounted for only around 10% of the total meat supply, had risen to the high level of 46% in 1947, two years after Japan's defeat. However, by the 1950s the figure had fallen back to around 30%. The increase can be explained, not so much by the fact that whale meat production had increased, but rather that it had come to occupy a greater proportion of the meat intake because of the big decline in the production of, or supply of, beef, pork and chicken as a result of the war. Conversely, as production of these farmed meats picked up, the proportion of whale meat began to decline again.

The influence of the war was felt particularly strongly in the pork and poultry industries. The number of pigs produced in 1946

Table 5.1: Developments in the whaling industry with particular reference to Japan

Ancient Period	Utilisation of drifting and stranded whales
Medieval Period	Harpoon Whaling by whaling groups
Latter half of 17th Century	Beginning of net whaling by whaling groups (whales driven into nets, killed with hapoons and processed)
First half of 19th Century	Discovery of the 'Japan grounds' by American sail-powered whaling ships. American-style whaling fleets enter Japanese coastal waters.
Latter half of 19th Century	Trial introduction of American style whaling (in which small boats, based on a large-scale sailing ship, used whaling guns in their pursuit of whales).
1878	Ōseminagare (The great right whale shipwreck). The destruction of the Taiji whaling group.
1897	Establishment of En'yō Hogei K. K. and Nagasaki Hogei K. K. Introduction of Norwegian-style whaling (whales taken with a harpoon with an explosive head, fired from a gun mounted on the bow of a steam-powered vessel).
1904	Russo-Japanese War. Japanese whaling companies monopolise whaling on the Korean peninsula.
1911	The Tōyō Hogei K. K. Same station riot incident
1934	Nihon Hogei K. K. becomes first Japanese company to engage in factory-ship whaling in the Antarctic Ocean.
1937	Conclusion of the International Whaling Agreement Incorporating a ban on taking right whales and grey whales.
1941	Last year of Japanese factory-ship whaling in the Antarctic Ocean. Factory ships conscripted to war effort.
1946	Reopening of Japanese factory-ship whaling in the Antarctic Ocean. Conclusion of International Convention for the Regulation of Whaling – 15 nations join.
1949	Inaugural general meeting of the IWC (formed in 1948). Total catch for the Antarctic set at 16,000 BWU (Blue Whale Unit: 1 blue whale = 2 fin whales = 2.5 humpback whales = 6 sei whales).
1951	Japan joins the IWC.
1962	Whaling quotas based on national allocations introduced in the 1962–63 season. On the advice of scientists (the so-called Special Committee of 'Three Wise Men') the catch for the 1963–64 season reduced to 10,000 BWU.
1966	Whaling ban introduced for blue whales and humpback whales.
1967	Trial Japanese whaling of minke whales in the Antarctic Ocean in the 1967–68 season.
1972	U.S.A. proposal for a 10-year total ban on commercial whaling, put to the U.N. Conference on the Human Environment, accepted. IWC rejects the proposal. But in the 1972–73 season BWU system abandoned and quotas for each whale species introduced.
1976	Ban on whaling of fin whales.

Table 5.1: continued

1979	U.S.A. enacts the Packwood Magnuson Amendment (within America's 200 nautical mile territorial waters, limiting the catch of countries in breach of the regulations of the International Convention for the Regulation of Whaling to 50% of their allocation in the first year and zero in the second year) (hereafter abbreviated to the PM Amendment). Ban on whaling of sei whales.
1982	IWC decides on a moratorium of commercial whaling, from 1986 for coastal whaling and from the 1985–86 season for factory-ship whaling. Japan registers its objection.
1984	America–Japan Whaling Agreement. Japan withdraws its objection and agrees to suspend commercial whaling from 1988 (1987–88 season). In return, the U.S.A. agrees to permit Japanese commercial whaling for a limited period of two years and not to impose sanctions under the PM Amendment.
1986	Japan claims its small-type coastal whaling belongs to the category of aboriginal subsistence whaling and requests a quota under this designation.
1987	Japan presents its plan for scientific whaling to the IWC. IWC recommends suspension of scientific whaling.
1988	Japan carries out scientific whaling in the Antarctic Ocean from the end of 1987. U.S.A. invokes the PM Amendment.
1992	Iceland withdraws from the IWC. Norway announces it will recommence commercial whaling from 1993.
1994	The IWC adopts a resolution for a sanctuary giving effect to a ban on commercial whaling in the Antarctic Ocean.
1997	Ireland proposes a compromise plan whereby, after the completion of the RMS, coastal whaling is permitted, but whaling in international waters, scientific whaling and international trade in whale meat is banned.
2002	Iceland permitted to rejoin the IWC reserving its opposition to the moratorium on commercial whaling.

was 88,082, less than one tenth of the peak of 1,140,479 reached in 1938.

It was not until 1956 that this pre-war figure was finally overtaken (Nōsei Chōsa Iinkai 1977: 256–7, 622).[1] The reason for the downturn in production seems to have been that it was no longer possible to 'import' from Manchuria the corn and soybean oilseed dregs used to make the concentrate fed to pigs and chickens in the home islands of Japan (Shiryō Haikyū K.K. Chōsaka 1943: 15, 129–34, 170–6; Matsuo 1964: 178–85, 226–7). On the other hand, the number of cattle, which can be fed on roughage (hay, straw etc), remained relatively constant even in the period around 1945, fluctuating at

Figure 5.1: Changes in Norwegian-style and factory-ship whaling in Japan (1945–1972)

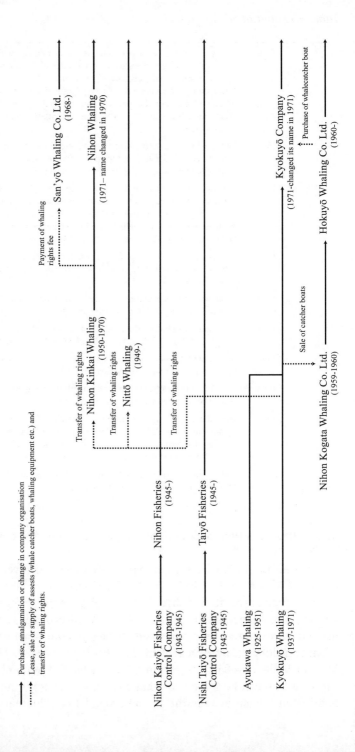

Figure 5.2: *Annual meat production per Japanese citizen*

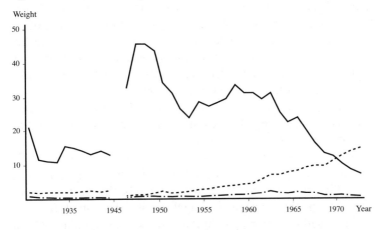

—— Whale meat as a proportion of total meat production (%)
----- Meat (beef, pork, chicken, other, whale) (kg)
—·— Whale only (kg)

Notes:
1. Figures for the proportion of whale meat are rounded to the nearest percent. (Translator's note: These statistics are for meat production (i.e. meat supplied to the market) rather than for actual consumption).
2. This graph has been made with data provided by the Agricultural Policy Survey Committee. Nagasaki Fukuzō (Nagasaki 1984) also includes the same data, but there are discrepancies in the figures for the overall meat supply from 1965 to 1972 and whale meat only for 1970. Further, there are several errors in Nagasaki's calculations of the proportion of whale meat in the total meat supply. Consequently, I have not used the data published in Nagasaki's article.
3. The sources used, do not clearly indicate what constitutes 'a Japanese' (*kokumin*). For the sake of statistics, it seems likely that the term does not include colonial subjects for the period up to 1945 and, post-World War II, Okinawans were not counted until 1972 and Amami Islanders until 1953 (Nōsei Chōsa Iinkai ed. 1977: iii). Similarly, we can surmise that the inhabitants of the Ogasawara (Bonin) islands were not 'Japanese' for statistical purposes for the period from the end of the war until 1968. Moreover, it is quite possible that the Okinawans, Amami and Ogasawara islanders were not considered 'Japanese' in the statistical sense even in the pre-war period.

Sources: Nagasaki, Fukuzō (1984), 'Nihon no engan hogei,' *Geiken tsūshin*, 355: 82; Nōsei Chōsa Iinkai ed. (1977), *Kaitei Nihon nōgyō kiso tōkei*, Nōrin Tōkei Kyōkai: iii, 346–7.

between 2 million and 2.5 million head. But the annual net food supply of beef per capita for 'the Japanese,' which had been fixed at between 700 and 900 grams per annum between 1930 and 1939, fell to 400 grams in 1946 (Nōsei Chōsa Iinkai 1977: 256–9, 346–7, 622). The reason for this is not clear, but we can surmise that perhaps it was because the imported beef that had accounted for from fifteen to thirty per cent of beef consumption in the home islands (Shiryō Haikyū K.K. Chōsaka 1943: 270–83) from the beginning of the 1920s was no longer available because of the war, or that the decline in incomes caused a temporary drop in demand for meat products (Matsuo 1964: 226–7). Be that as it may, to return to whale meat, it is a fact that this food source was never any more than a supplement to fish, beef and pork and that the years 1950 and 1951 actually witnessed such sluggish sales of whale meat that it was sold off cheaply for school lunches and the like.

1951 was the year in which Japan became a signatory to the International Convention for the Regulation of Whaling. By 1955 Taiyō Fisheries' *Nisshin Maru* had come to be the most successful vessel among the nineteen whaling fleets of all six nations involved. But as it had become apparent that stocks in the Southern Ocean were being devastated, the IWC quotas were gradually reduced from 15,500 BWU in the 1953–54 season, to 15,000 BWU in the 1955–56 season and 14,500 BWU (raised again to 15,000 BWU after objections by all the operating nations) in the 1956–57 season (Kawashima and Katō Hidehiro 1991). But reduction in the quotas meant that whaling became less viable economically, so the 'whaling Olympics' was abandoned and there was an attempt to introduce national quotas. But the talks broke down without reaching agreement, when the various whaling nations became engaged in a game of tactics in which each sought to secure an advantage in the allocation of quotas. Consequently, from the 1959–60 season to the 1961–62 season, whaling was carried on under the 'self declaration' system, in which each nation reported its whaling quotas in BWUs (Kawashima and Katō Hidehiro 1991). Japan had five fleets in 1956 (one of the factory ships had been purchased from Panama by Kyokuyō Whaling). In 1957 this grew to six fleets when Taiyō Fisheries bought a factory ship from South Africa, and increased to seven fleets in 1960 when Kyokuyō Whaling purchased a factory ship from Britain. Then in the 1960–61 season Japan declared its self-imposed quotas at 5,980BWU.

It was around this time that the IWC finally began carrying out serious research. Later, this task was taken over by a group of scientists known as the Special Committee of 'Three Wise Men.' This committee, established in 1960 (Ōmura 1963: 24–6), advised the IWC Scientific Committee in 1963 (Morita 1963b: 30–1) that; 1. an eight-year ban on whaling would be necessary to restore stocks to their 1956 levels, and 2. to maintain the *status quo*, for the time being, the annual catch should be kept at under one quarter of the current limit of 15,000BWU (Hara 1993: 111–12). From around the time this advice was issued, the IWC began to become inclined towards conservation. First, from the 1962–63 season the national allocations were put in place (Japan 33%, Norway 32%, the Soviet Union 20%, Britain 9% and the Netherlands 6%. Japan's share increased to 41% when it purchased two additional factory ships, one from Norway and one from Britain, each with a 4% quota) (Kawashima and Katō Hidehiro 1991; Morita 1963a: 30–1). Subsequently, on the advice of the Special Committee of 'Three Wise Men,' whaling quotas were cut in quick succession to 10,000 BWU in the 1963–64 season, 8,000 BWU in the 1964–65 season (this figure the result of negotiations among the operating nations) and to 4,500 BWU in the 1965–66 season. In addition, a ban was imposed on the taking of humpback whales in the Southern Ocean from the season of 1963–64 and the ban was extended to blue whales in the 1964–65 season (Kawashima and Katō Hidehiro 1991). In response to these actions of the IWC, Japan decided to increase its own whaling quota by purchasing factory ships from those nations that had abandoned whaling because it was no longer profitable. Even so, Japan was compelled to reduce the number of its whaling fleets to five in the 1965–66 season (Taiyō Fisheries and Kyokuyō Whaling each cutting one factory ship) (Morita 1965: 20–1) when its quota was set at 2,340 BWU, only about 36% of the 6,574.13 BWU of the 1961–62 season, the zenith of Japanese whaling in the Antarctic Ocean (Kawashima and Katō Hidehiro 1991). In other words, in spite of the devastation of whale stocks in the Antarctic and the inevitability of reduced whaling quotas, Japan continued to adopt an expansionist line that ultimately was to lead to its downfall.

Within the IWC, the confrontation between the whaling nations and the non-whaling nations, which were calling for conservation, became even more determined. But in reality from the 1964–65 season onward it was mainly sei whales that were taken (Kawashima and Katō Hidehiro 1991). And from the 1967–68 season, on a trial basis, Japan began catching minke whales, a species in which it had previously shown little interest (Kawashima and Katō Hidehiro 1991). In the 1971–72 season, apart from the Norwegian quota of 50 BWU, the quota was divided up in proportion to the number of whaling fleets, between Japan (1,346 BWU) and the Soviet Union (904 BWU) (Kawashima and Katō Hidehiro 1991). In this year (1972) the United Nations Conference on the Human Environment, meeting in Stockholm, adopted an American proposal suggesting a ten-year complete ban on commercial whaling. The IWC rejected this advice, but abolished the BWU system from the 1972–73 season, and adopted quotas for each whale species (Kawashima and Katō Hidehiro 1991).

Let us now consider the situation with coastal whaling. Immediately after the end of World War II, the distribution of coastal whaling catcher boats saw Nippon Fisheries with nineteen vessels, Taiyō Fisheries with five and Kyokuyō Whaling with one. However, when the Law for the Elimination of Excessive Concentrations of Economic Power, proclaimed in 1947, was applied to Nippon Fisheries, the company was compelled to transfer the whaling rights of three vessels to Kyokuyō Whaling in 1948. Further, in 1949 the Nittō Whaling Company was established with whaling rights (for one vessel) transferred from Nippon Fisheries under the same law and in 1950 Nihon Kinkai Whaling Company was established with rights for two whaling vessels from Nippon Fisheries.[2] In addition, small type whaling, that had been unrestricted in pre-war years, came to require regulation, in the form of a permit from the Minister of Agriculture and Forestry, due to the flood of entrepreneurs entering the industry on the wave of the post-war need for increased food production (Maeda and Teraoka Yoshirō 1952: 117).[3]

From around 1950 the eastern Hokkaidō seas became the main coastal whaling ground and all the companies concentrated their activities in these waters. The Fisheries Agency established a quota of 2,400 sperm whales in 1956, reducing it to 2,100 in 1959 and 1,800 in 1962 (Watase 1965: 40), in a trade-off for increased quotas for factory-ship whaling in the North Pacific. On paper, whaling within the sperm whale quota system followed the pattern of 'the whaling Olympics,' but in practice publicly announced estimates of the catch for each catcher boat were fixed by a conference of the five whaling companies. In 1963, in accordance with the whale catcher regulation policy, which recognised larger vessels to compensate for retired tonnage, Nippon Fisheries cut three more catcher boats from its licenses, leaving a total of eight ships (Watase 1965: 41). In addition, the Fisheries Agency not only cut one fleet from Kyokuyō Whaling's Antarctic operation in 1965, but also closed down the company's unprofitable whaling station on South Georgia Island,[4] and in return for revoking the whaling rights of all five ships engaged in similarly unprofitable coastal whaling operations, implemented interim measures from 1965 to increase Kyokuyō Whaling's quota for factory-ship whaling in the North Pacific to 200 BWU (Morita 1965: 25–6). Thus, from around this time, were the larger enterprises among the whaling companies compelled to reduce their coastal whaling activities.

On the other hand, it is reported that as at April 1952 there were seventy-five catcher boats licensed to small-type coastal whaling throughout the country (Maeda and Teraoka Yoshirō 1952: 118), but excessive competition had thrown the industry into an economic slump. In response, in 1957 the Fisheries Agency decided on the policy of restricting to within about twenty-five vessels the permitted number of catcher boats engaged in small-type coastal whaling. Further, the permissible increased tonnage of catcher boats engaged in large-type whaling, was limited to the amount of tonnage of large-type or small-type whaling catcher boats that had been taken out of service. In addition, the agency permitted small-type whaling operations to participate in factory-ship whaling operations targeting sperm whales in the North Pacific, on the condition that they give up the small-type whaling for which they held permits and engage in new cooperative ventures ('Meidō shita hogei gyōkai' 1957: 22–25). So, in 1957 a number of coastal whaling operations pooled their 375 tons (it is unclear what type of ton is specified here (Kondō 2001: 336–7)) worth of whaling rights, formed a union and chartered a whale catcher to participate in a fleet. Later, in 1959, the union became the Nihon Kogata Hogei Yūgengaisha (Nihon Small-Type Whaling Company Limited).

By the beginning of the 1970s it had become obvious that sperm whale numbers in the eastern Hokkaidō seas had dwindled markedly and subsequent whaling operations came to focus on the waters off Kinkazan Island in Miyagi prefecture. Moreover, although the quota on sperm whales had been lifted in 1967 at the request of the four companies involved in coastal whaling, limits on catches of sperm whales were reimposed from 1969 when the United States, Japan, Canada and the Soviet Union agreed to place quotas on whale catches in the North Pacific, both for factory-ship whaling and land-based whaling. Besides, after 1965 the taking of blue whales and humpback whales in the North Pacific was prohibited (Nihon Honyūrui Gakkai 1997: 175–6, 178). On the other hand, San'yō Hogei Yūgengaisha (San'yō Whaling Company Limited) had been established in 1968 as a result of the policy that permitted small-type whaling operations to reduce their catcher boats and change over to large-type coastal whaling. It is clear from these developments that coastal whaling was in a marked state of decline. The number of catcher boats involved in large-type whaling in 1972 had dropped to twelve; three each for Nippon Fisheries, Taiyō Fisheries and

Nittō Whaling, two for Nihon Whaling (Nihon Hogei K. K., Nihon Coastal Whaling having changed its name in 1970) and one for San'yō Hogei Yūgengaisha (San'yō Whaling Co. Ltd). Further, from 1968 the official number of small-type whaling catcher boats was pegged at ten vessels (Nagasaki 1984: 80).

To this point we have been considering developments in the Japanese whaling industry from the end of the Second World War to 1972. While acknowledging the need for more detailed chronological divisions, along the lines of my classification of the stages of development of whaling in the early modern period in Chapter 1, it is rather difficult to classify the development of Japanese whaling overall. This is because, in the period under question, Japanese whaling operations covered the Antarctic Ocean, the North Pacific and the Japanese coast and took the forms of factory-ship whaling, large-type and small-type whaling. If we were to attempt a division it might possibly be, first, from 1945, following the end of the confusion arising from the war to the eventual beginnings of an international framework when Japan joined the IWC in 1951. Then followed the period of rampant overfishing from 1952 to 1965 and finally, from 1966 to 1972, the period when the brakes began to be applied to indiscriminate depletion of whaling stocks.

Whaling discourse to the end of WW II

Hōshūtei's 'sense of whaling'

Next, let us consider, the narratives of those involved in whaling, in the light of what I have said regarding the development of Japanese whaling in the early modern period and just now in the post-war period. In this section, I shall be concentrating on the analysis of whaling discourse in the early modern period, but first I would like to mention a little about the way those engaged in net whaling felt about catching whales. There are very few sources that can enlighten us on this point, but there is one that gives us a relatively clear account of the situation. That is *Ogawajima keigei gassen* (The battle with the whales at Ogawajima) written in 1840 by one Hōshūtei Riyū (Hōshūtei 1840).

This is an illustrated book depicting whaling as it was carried out by a whaling group of Yobukoura bay in the Karatsu domain of the province of Hizen (present day Yobukochō, Karatsu city, Saga prefecture). It is not clear who Hōshūtei Riyū was, but, to judge from the fact that he has included Chinese and Japanese poems here and there in the text, we can surmise that he was, in his day, a man of letters (Tajima 1995: 386, 388). As the book is thought to have been presented to a local shrine by the manager of a whaling group, it is considered possible that the manager of the group may have had the book written expressly as a votive offering to the shrine. For this reason, I think we can regard the text as being not merely a depiction of events, but also, to a small extent, a reflection of the thinking of those involved in whaling.

When we look at the content of the book, we are struck by the fact that, as the title suggests, it depicts whales as enemies and the members of the whaling group as if they were soldiers challenging a foe in battle. In concrete terms, the foes in this case are a right whale and its calf. While today there are problems associated with the taking of whales with calves, for the net whaling groups, parent and calf together made an ideal target. This we know from the words of songs sung at the celebratory banquets of whaling groups (Kizaki 1970: 781–82; Yoshioka 1938: 41–4 (1973b: 465–8)) and Morita Katsuaki has also pointed this out in his research (Morita 1994: 171–80).

In so much as it depicts the sequence of activities of the whaling group associated with the capture and butchering of whales, this work is not so different from others portraying the work of whaling groups. But, in addition to the depiction of the hunt as a battle scene, as we have observed above, Hōshūtei's book is characterised by what may be considered his consciousness of whale hunting and the role of the *hazashi* (See Chapter 1). This is particularly clearly demonstrated in the final section of the book.

In this passage, after citing examples of a whale's devotion to its calf and how, in its final moments, the whale turns to die facing west [Translator's note: the direction of the Buddhist paradise], (from our modern perspective the veracity of this statement is questionable), he goes on to say what a terrible thing it is to have no regrets over making a living by killing such compassionate creatures and then frittering away the money thus earned on frivolities. He continues as follows.

People of old said that once you had heard that cry you could not bear to eat their flesh. How merciless it is to feel no pity for that resounding cry of pain as they face west to die, then row the carcass in to shore, cut it up in the barn and then immediately boil the meat or grill it on a metal hot plate and smack your lips as you savour the taste. As

someone was relating this with furrowed brow, a man came up beside
him and said in reply 'What you say is so, but for man too there is life
and death. It is the same with all things. The life of a great whale and
that of a tiny whitebait are no different. In the past there were whaling
groups throughout the land, (section omitted) if the whales were not
taken here they would be killed somewhere else. We have no say over
life and death. The time comes to boil the whitebait in the cooking pot.
Thus, time comes for everything. It should in no way be regarded as a
sin, since the killing is not without purpose. At times a life cast away in
the shallows can bring benefits. A dead whale is worth a huge amount
of money. The meat of that whale may be savoured and appreciated in
the mouths of several thousand people; no part of the hide or flesh of
such a huge 'fish' is discarded, it helps maintain the lives of several
hundred people and enables them to offer much tribute to the lord of
the domain. Even those of neighbouring districts, hamlets, inlets and
islands, including even children and women benefit from its blessings.
Such are the great merits of whales' (Hōshūtei 1840: 361–2).

Hōshūtei goes on to record how, after the completion of the hunt,
a service for the repose of the soul of the dead whale, in which
the *hazashi* and others involved participate, is held at the local
temple.

What we first notice in this description by Hōshūtei is that there
was a consciousness among the people of the day that it was heartless
to kill and make use of whales. Secondly, we are struck by the fact
that, the arguments advanced today by the proponents of whaling,
namely, that whales as living organisms are no different from any
other and that all parts of the whale carcass are utilized, were, even
at this time, already being made to counter expressions of that
consciousness. But in spite of these, in a sense, rational arguments,
and the assertion that 'it was not a sin,' they cannot escape the fact
that there was a certain sense of guilt over killing whales. This
explains the performance of the ritual of the solemn Buddhist
memorial service. I think this state of consciousness which falls
outside the realm of rational argument has important implications
for our consideration of future relations between human beings and
whales. Nevertheless, the problem is that, even though memorial
services were held for whales, these were unable to stem the tide of
indiscriminate harvesting. Even during the blatant overfishing of
modern times the whaling companies carried out memorial services
for the whales they had taken and there were instances such as that
of the harpoon gunner who dedicated a temple bell in appeasement

for the whales he had killed (for example, Nasu 1989: 6). Of course those conducting, or participating in, these rituals probably did so with a range of thoughts in mind. But I think it is at least possible to posit a change from the schema {guilt over killing whales} → {memorial service} to the schema {performance of memorial services} → {killing any number of whales with impunity}.

Oka Jūrō's inexhaustible supply theory

Let us now turn to the analysis of the discourse of those involved in whaling in the early modern period. First, I would like to consider a figure who has already appeared in Chapters 1 and 2 of this book, Oka Jūrō, the president of Tōyō Whaling.

In *Honpō no Noeruēshiki hogeishi* (A history of Norwegian-style whaling in Japan) edited by the Tōyō Whaling Company, a source I have already cited many times, there is a record of a conversation with Oka Jūrō (Tōyō Hogei K.K. 1910: 1–36). This is thought to have been made at a meeting held on 13 January 1910 at the Japanese restaurant *Nadaman* in the Kitahama district of Osaka. Present at the meeting were the directors of the Kitahama Bank, the directors and a member of the futures and spots brokerage committee of the Osaka Stock Exchange, and the directors of Tōyō Whaling. That is to say, this speech by Oka Jūrō was made directly after the amalgamation of the whaling companies (See Chapter 1) in front of representatives of the banking and securities industry, so we need to consider carefully what the intention of his statement might have been.

After giving a concise description of hunting and processing activities of Norwegian-style whaling, the process of the introduction of Norwegian-style whaling into Japan and the details of the amalgamation of whaling companies, Oka goes on to explain the structure and commercial outcomes of Tōyō Whaling. He follows this up by addressing the suspicion that some might feel that if 'such a large-scale organisation' (Tōyō Hogei K.K. 1910: 28) were to engage in whaling every year it might lead to the extermination of whales. Over several pages, he replies in 'rather specialist terms from the point of view of a whaler' (Tōyō Hogei K.K. 1910: 28).

At this point Oka details two theories regarding the future of harvesting, not only whales, but marine resources in general. One of these, 'the theory of extermination through ceased propagation' (Tōyō Hogei K.K. 1910: 28), he explained, holds that repeated whaling over time leads to progressively smaller animals being taken and this results in the 'mother fish' (*bogyo*) (Tōyō Hogei K.K. 1910: 29) with reproductive capacity being fished out, eventually causing the extinction of the species. The other theory, 'the inexhaustible supply theory' (Tōyō Hogei K.K. 1910: 28), claims that as long as a marine organism's source of food is assured in a certain fishing ground, that fishing ground will remain viable indefinitely, because even if, for example, the 'mother fish' in that fishing ground are fished out, others of the same species will come in from other areas to fill the void (Tōyō Hogei K.K. 1910: 28–9). Oka then goes on to say that, unlike lakes and rivers, the ocean is vast and obviously fish can move to areas where food is plentiful. And with the growth of the fishing industry and the increase in the number of fishing boats, while there is a reduction in the catch per vessel, the total catch does not decrease but increases. Oka says he supports the inexhaustible supply theory when it comes to whaling because of the abundance of the whales' food, like krill and squid, in the waters around Japan and the Korean Peninsula (Tōyō Hogei K.K. 1910: 29–32).

On the other hand, Oka also mentions that you rarely see right whales and that the catches of humpback and blue whales are decreasing every year (Tōyō Hogei K.K. 1910: 32–3). This he puts down to the fact that whales have the ability to sense danger and that they will leave a fishing ground if they are chased by catcher boats. He goes on to explain that while it may appear that the absence of right, humpback and blue whales means that a fishing ground is no longer viable, this is because these species are comparatively sensitive to signs of danger and are slow to move back into a fishing ground, but that provided ocean currents have not altered the whales' food supply, there is no need to feel pessimistic. He claims that, actually, grey whales and fin whales frequently move back into fishing grounds and there has been no reduction in the catches of these species (Tōyō Hogei K.K. 1910: 32–4).

There can be little doubt that Oka's highly optimistic view on the reduction of whale numbers, delivered to members of the banking and securities sector, reflects his desire to attract investment and support from this quarter by talking up the prospects for growth in the whaling industry at this time, ten years after he himself had established Nihon En'yō Fisheries. We can also assume that he had no choice but to support the inexhaustible supply theory as the theory of extermination through ceased propagation, which implied regulations limiting catches, was problematical for the economic activity of using whales, as I mentioned in Chapter 3. But even beyond these considerations, we cannot deny the possibility that Oka himself was speaking out of the conviction that the inexhaustible supply theory was valid.

Either way, Japanese whaling in the early modern period began with the application of this kind of intensional logic. You may wonder what form the narrative took later, when Norwegian-style whaling expanded beyond Japan to the Antarctic Ocean and the reduction of whale numbers was beginning to become obvious? To answer this question I would like to turn to the description of another individual.

Baba Komao's monopoly argument

Here, I would like to consider Baba Komao's book *Hogei* (Whaling) (Baba 1942), which I used as source material in Chapter 1, mainly in the discussion of the introduction of factory-ship whaling. Baba's book says little about his background, but he was in charge of the Nihon Whaling Company's first whaling project in the Antarctic Ocean where he directed operations from on board the *Antākuchikku Maru* and later spent a further four seasons engaged in factory-ship whaling in the Antarctic Ocean (Baba 1942: 96). We can assume from this that he was at the time in a management position in the whaling company.

In his book of over three hundred pages, Baba gives a comparatively detailed account of all aspects of whaling and in his final chapter, Chapter 7, he deals with 'the problem of whale conservation and limits on catches' (Baba 1942: 301–326). Here he first gives an overview of the regulations in Japan, then goes on to consider the implications of the International Whaling Agreement (See Chapter 1). Then, he concludes with a section entitled 'the significance of whaling management' (Baba 1942: 321) in which he states that the two most common harmful effects of the boom in whaling 'are indiscriminate fishing' (Baba 1942: 321) and obstruction of other industries. As an example of the former, he cites the situation before the birth of Tōyō Whaling, when, according to Baba, the establishment in rapid succession of a large number of whaling companies led to overproduction and a sudden fall in the price of the industry's products. Competition and the depletion of fishing grounds resulted in greatly increased production costs. Consequently, the running of whaling companies become precarious financially and there was a marked decrease in the numbers of migrating whales. In response to this situation, he writes, the Regulations for the Management of Whaling were issued and have provided a continued and stable whaling environment ever since. His example of the latter harmful effect is taken from Norway. Baba recounts how the rapid development of coastal whaling in Norway towards the end of the nineteenth century, severely damaged cod and herring fishing, which had previously been the most important industries, and gave rise to such a violent protest movement against whaling that the government was compelled to enforce successive restrictions and eventually a ten-year prohibition on coastal whaling (Baba 1942: 301–326).[5]

But, Baba continues, these examples were very much dependent on the particular circumstances of each of the countries involved and are very different from the situation in the open, international waters of the Antarctic Ocean, where the usual territorial rights do not apply. Further, he agrees that it was absolutely right that, with the expansion of factory-ship whaling in the Antarctic, the call for protection of whales and the speedy enforcement of restrictions should be raised, given the experience of the past when whales may well have been driven to extinction if the situation were permitted to continue unchecked. But, he says, 'the managers of businesses want those enterprises to which they have devoted themselves to be prosperous. And countries naturally hope the businesses of their nationals will flourish. Moreover, this desire is accompanied by a considerable sense of exclusivity' (Baba 1942: 323). Consequently, 'It follows that when results indicate a business is in danger of collapse, there is a desire to constrain the operation to a necessary extent in order to ensure the continuity of the business itself while attempting to place oneself in the most advantageous position possible' (Baba 1942: 323).

He then goes on to say that although the International Whaling Agreement has been achieved, no firm international consensus has been reached on limits to the number of vessels or the amount of product. Baba proceeds to point out that 'those who advocate restrictions' (*seigen ronja*) (Baba 1942: 323) do so for one of two reasons, either on economic grounds or 'from a position of conservation of rare animals, a view promoted mainly by biologists' (Baba 1942: 324). The former group argues that reduced whale stocks brought about by unrestricted whaling will result in greatly increased production costs and overproduction that will cause confusion in the whale products market. So, they claim, appropriate limits should be applied to whale catches in order to ensure the long-term continuity of the business. In addition, they argue that, at this present stage, 'the utilisation of the precious whale resource' (Baba 1942: 323) is still inadequate, but as it is expected greater use will be made of it in the future, the 'resource' should be protected until it has been used. The latter group claims whales are among the very few 'relics of a former age' (Baba 1942: 324) still living on the earth today and although they are ultimately destined for extinction, they are special, extremely precious and biologically unique animals that have developed over hundreds of thousands of years. Yet in spite of this, and even though their contribution is minute in comparison with all the 'resources' of the earth, it is 'a shameful outrage that human beings should violently drive these animals to extinction' (Baba 1942: 324) simply because we can make use of them and they bring profit to a very small number of whaling companies (Baba 1942: 323–4).

A variety of arguments were put forward in response to this 'extinction of whales theory' (*geizoku senmetsu ron*) (Baba 1942: 324), in so far as it applied to the regulation of whaling, from those in the industry who were in the 'anti-restriction' (*seigen hantai ronja*) (Baba 1942: 324) or 'non-restriction' (*hi seigen ronja*) (Baba 1942: 325) camps. One was the assertion that since the industry required high capital and running costs, whaling would automatically come to a stop when income fell below the level of costs. And the borderline between profit and loss would be reached far earlier than the point of extinction of whales. Moreover, the Antarctic Ocean is the whales' feeding and breeding ground. The number of whales taken there each year is only a small proportion of the total; the majority survive to go on breeding, so fears of extinction are groundless. The more extreme voices in the anti-regulation lobby say, 'even if, for the moment we concede a little ground, ultimately the time of extinction will come. It is natural in the scheme of things for mankind to use not only the whales, but also all the blessings of nature for its own benefit. Indeed, it is our duty to do our utmost to direct all the resources of the human intellect to this task. We should take as many natural products as possible and apply them to the welfare of mankind' (Baba 1942: 324–5). Baba points out that these people assert that the argument put forward by some of the advocates of restrictions, namely, that a moratorium on whaling should be enforced until an effective method of utilisation has been found, is just a meaningless attempt to buy time, which pays no heed to the realities of the situation (Baba 1942: 324–5).

Nevertheless, as Baba says, even the advocates of restrictions were not demanding a complete ban and the anti-restriction lobby wanted to see a continuation of the industry under appropriate restrictions, so the International Whaling Conference was able to achieve a degree of success with the majority supporting the argument for restrictions. The main problem, however, as Baba is at pains to point out, is the clash over the national interests of the nations involved. In particular, those countries not engaged in whaling and with no market for whaling products, attending the International Whaling Conference simply because they might one day be involved in whaling, abused their welcome by putting forward 'irresponsible arguments for restrictions based on lofty speculation' (Baba 1942: 325). He goes on to say that it is inevitable, given the nature of international conferences, that one country's profit means another country's sacrifice, but it is regrettable when some countries use this as a reason to refuse to attend the conference (Baba 1942: 325).

What, given the circumstances of the International Whaling Conference, did Baba himself think of the debate? He concludes his account with the following statement.

> For our part, if one day the world whaling industry were to become unprofitable because of the reduction of whale stocks, and at that level Japanese whaling could keep a little in reserve and stand alone, it would make international whaling a Japanese monopoly. It is only then, that it would be possible to have optimally rational restrictions. Moreover, achieving this position is not simply a dream. I am convinced that it is entirely possible and we Japanese whalers have the heavy responsibility of making it a reality (Baba 1942: 325–6).

Above we have considered Baba's narrative in some detail. What emerges clearly from his account is that the argument over protection of whales is beginning to become an issue. At this stage the focus is ultimately on the two aspects of importance to science and economic viability. These are different from the reasons for the protection of finless porpoises covered in Chapter 3 and, understandably, they are not based on ecological concepts. Further, we note that, at this time, some of those opposing these views took the extreme position that it is natural for human beings to make as much use of nature as they can and if that meant the extinction of whales, then so be it. Baba's own position in the debate seems to have been that he believed whaling would automatically cease at the level when it became unprofitable and that this would happen long before whales became extinct.

Taking this position as his premise, Baba feels that in the unavoidable rivalry of countries pursuing their own interests, if Japan were able to gain an upper hand in the competition between nations, those nations whose interests were not being met would withdraw, to the point where finally whaling would become a Japanese monopoly. If that were to happen, he thought, regulations could be introduced without consideration of the interests of other nations, in the same way as restrictions had been introduced within Japan after the establishment of Tōyō Whaling.

Baba's ideas were very much a product of the extreme ethnocentrism that characterised the period directly after the outbreak of the Pacific War. And he is extremely optimistic in his assertion that before whales were wiped out, whaling would automatically come to a halt when it reached a level where it was no longer profitable, i.e. that even if competition continued to the point where many countries were driven out of the industry, whale stocks would never be reduced to the level of extinction. Let us now turn to the post World War II whaling discourse to determine whether these ideas of Baba, and Oka before him, were confined to the pre-war period.

Post World War II whaling discourse – the significance of 'what cannot be said'

Here too, it might be seen as appropriate to follow the method of the preceding sections and consider the discourse of those connected with whaling in the period after the Second World War. However, in going through the materials related to fishing activity after World War II, I was unable to turn up much, of the kind of material we considered in the previous section, on ideas about whaling or clear developments in response to diminishing whale numbers. In particular, I searched through all the articles, from 1945 to 1973 (Nos 745 to 1068), with any conceivable connection to whaling, in *Suisankai* (Fisheries World), the representative trade journal of the Japanese fisheries industry but the result was much the same.

Perhaps this fact can simply be summed up in the phrase 'lack of sources,' but I would like to explore the matter a little more deeply from the perspective of 'things that cannot be said.'

It is possible, I suppose, that what could not be said was something other than the fact that whale numbers were decreasing or that there was overfishing of whales. But given the post-war system of whaling we examined in section two of this chapter and the pre-war discourse on whaling in section three of this chapter, we can say that that interpretation is hardly possible. In which case, we are left to conclude that one could not refer to, or talk about, the fact that whales were decreasing, or that they were being hunted indiscriminately.

We can say that this hypothesis is not necessarily an exaggeration. We know, because in recent years it has come to light, from among those involved in whaling, that the Japanese whaling industry manipulated whaling statistics. In short, they cooked the books in regard to the headcount of whales (Kondō 2001: 339–42, 401–11; Watase 1995). According to Kondō, from around 1950 the facts of the harvesting of sperm whales off the Japanese coast began to be concealed. At this time the numbers of whales concealed were very small, when undersized animals were included in the total catch and the number of whales recorded was adjusted against the volume of product produced to make it appear as if all were of a legitimate length. But, the concealment really started in earnest when, as I mentioned before, quotas on sperm whales began to be introduced from 1956. This was because the quotas allocated to each of the whaling companies were insufficient to be economically viable. According to the materials Kondō has made public, one company that had a quota of 924 whales for the 1970 season actually caught 1,410 whales in that year. Similarly, in 1971, when its quota was 825 whales, it caught 1,336 and in 1972 it caught 1,243 whales when its quota was actually 666 (Kondō 2001: 408). Moreover, Kondō and Watase Sadao point out how inspectors from the Fisheries Agency were entertained at restaurants and hot spring resorts to keep them away from whaling stations while undersized whales were being flensed.

As this was the practice of the day, Kondō asserts it is meaningless to simply list the published statistics of the number of sperm whales taken after 1950 (Kondō 2001: 405).[6]

Similar concealment was, according to Watase, also going on in the Antarctic Ocean. Soon after the war, when the processing capacity of whaling factory ships was low, there was a system in place under which each of the catcher boats assigned to a factory ship was allocated a quota of whales, and a commission was paid to those boats according to the size of the whales caught. So, it became important to catch larger whales within one's quota. Consequently, at the end of a day's hunting, the larger whales would be taken to the factory ship to make up the quota and any smaller whales caught would be deflated (See Chapter 1) and left to sink. According to Watase, the number of whales discarded in this way in a single whaling season was around fifty whales per boat, or between three hundred and three hundred and fifty for each whaling fleet. This means, at a conservative estimate, that a total of 3,000 whales were discarded in this way before the new factory ships were brought into service. He goes on to say that even from beginning to end of the quota system just mentioned, throughout the entire period of factory-ship whaling in the Antarctic Ocean, illegal catches of undersized whales and whales with calves were discarded and not delivered to the factory ship.

Given the foul play that was going on, we cannot but surmise that it was not possible to acknowledge overfishing and the reduction of whale numbers, or talk frankly about the kind of systematic restrictions needed to redress these problems. Kondō puts it like this, 'for those in the whaling industry, it was taboo to mention that whale resources were drying up. Faced with the reality of diminishing numbers of whales, the whalers put forward a variety of other possible reasons why this should be so, such as; changes in ocean currents, insufficient food supply, changes in water temperature, whales moving further out to sea, and so on.' Or,

> not only has the Japanese whaling industry never shown any interest in whale conservation, but it has also been slow to investigate the reasons for reduced catches. The whalers were only interested in profits and, even though the numbers of migrating whales were diminishing, they would never say that the resource was drying up. It was absolutely forbidden for those in the whaling fraternity to even mention the subject (Kondō 2001: 334, 416).

Nevertheless, even though there was no mention of the drop in whale numbers, we need to investigate whether anything was done to respond to this phenomenon and to interpret this as statements of those involved in whaling. As it happens, I have been able to turn up a little of this kind of material in the *Mainichi Shimbun* newspaper column, 'My opinion,' written by Miyata Takeshi, then managing director of Nippon Fisheries Co. (Miyata 1959).[7]

In his article, Miyata first points out that Japan, of all the countries in the world, makes the most complete use of the whale and goes on to give a number of examples. Then he continues to explain that, at the time, 15,000 BWU were being taken each year at roughly the same pace and there was no marked change in the size, or in other biological data, of the whales caught. This, he writes, 'seems to indicate that the present whaling quotas will more or less maintain the resource' (Miyata 1959). Then, after giving a brief explanation of the mechanisms of the International Convention for the Regulation of Whaling and the rules associated with it, Miyata goes on to conclude as follows.

> Recently there has been talk that whales are decreasing and they are becoming harder to catch. We cannot deny that whales have become cleverer at avoiding capture and are very difficult to catch. At present, with the free competition in the whaling industry, there are inevitably winners and losers. The losers come up with various economic reasons and excuses to avoid losing face, but the winners, for their part, have put in the mental and technical effort and the appropriate high level of investment that that entails. It might be a mixture of good and bad, but when it boils down to it, those with the greatest quantity win and it is unreasonable to ask them to settle for less. The meeting to allocate national quotas for whaling in the Antarctic Ocean is a shambles. I wish they could accept that the winners are good and deserve to be rewarded, rather than all insisting on blowing their own trumpets (Miyata 1959).

Here, is Miyata telling us *what he cannot tell*? Or is he perhaps completely refusing to consider what has been going on? At any rate, Miyata denies that whales are decreasing and wants to maintain the *status quo* for whaling quotas. Further, like Oka mentioned above, Miyata states that 'whales have become cleverer at avoiding capture.' And finally, in the section quoted above, Miyata seems to be saying that, as the *Nisshin Maru* of the Taiyō Fisheries Company emerged with the top results of all the fleets fishing in the Antarctic in 1955, the Japanese fleets had been 'good' because of the considerable effort and investment they had put in. By gradually increasing its fleets and its catch Japan had won out against its competitors, so the meeting to allocate national quotas for whaling in the Antarctic Ocean should also ensure that 'those with the greatest quantity' won. Although he is not going so far as to suggest that Japan be given a whaling monopoly, we can assume that Miyata, like Baba mentioned above, takes the view that Japanese whaling will come to occupy a dominant position internationally by continuing to promote Japan's interests in the competition with other nations as those countries unable to profit from whaling gradually withdraw from the race. Again, like Baba, he seems to take the optimistic position that even if competition continues, the number of whales will not decrease to a level where they are likely to become extinct. But Japanese whaling, gave little positive thought to the prevention of overfishing, rather, because some skulduggery was going on in the form of the concealment of catches, Miyata's opinion was just another case of someone blowing his own trumpet.

Summary

The logic of overfishing, as it emerges from the above discussion, can be said ultimately to have been have been shaped as a convenient rationale to increase profits in the whaling industry and to gain hegemony in the whaling world. Within that space, the fact that whale numbers were decreasing could not be mentioned. On the other hand, frequent use was made of the kind of argument Oka put forward in support of his inexhaustible supply theory, namely, that the apparent reduction in whale numbers was simply because whales were becoming more adept at avoiding capture. Further, the idea was that competition would continue on the basis of these optimistic observations and Japan would emerge as the victor. In fact, we can say that, in the Antarctic Ocean at least, Japan succeeded in achieving a dominant position in whaling from the beginning of the 1960s. But this was at the cost of considerable environmental destruction, to the point where the numbers of several species of whale decreased and finally there was no choice but to greatly reduce catches.

However, I wish to stress the point that I am not claiming that 'the logic of overfishing' I refer to here has been an unchanging essence of Japanese whaling throughout its history. To do so would be to engage in the very kind of substantialist argument I criticise in this book. Rather, we should think of the logic of overfishing as an idea that can appear at any time in any place.

What then can be done to escape from this logic of overfishing? A hint to the answer to this question emerges clearly, I think, in Hōshūtei's description of the guilt, or the state of consciousness, over the killing of whales, which falls outside the realm of rational argument. And the presence of this consciousness is not confined to the narrative of whaling in the Edo period.

As we can see from the words 'science reader for young Japanese' (*shōkokumin kagaku yomimono*) inscribed on the inside cover of *Hogeisen nikki* (Whaling ship diary) written in 1941 by Marukawa Hisatoshi, the book dealt with a number of topics on the science and technology of whaling, particularly whaling in the Antarctic

Ocean, written with simple explanations for children. It contains a description along the following lines (Marukawa 1941: 174–9).

In the treaty signed by members of the whaling industry from each country (presumably the International Whaling Agreement, see Chapters 1 and 3) it is agreed that suckling whale calves (*chinomi kujira*) (Marukawa 1941: 174) and mother whales accompanying suckling calves cannot be caught. But as Marukawa explains, in the past, large numbers of these whales were taken, as suckling calves swim slowly and mother whales adjust their own pace to that of the calf and stay on the surface to take care of the calf rather than diving and surfacing at will as they normally would. In this way, as was the case with the whaling groups, a whale with a calf was regarded as the most suitable target. The calf would be taken first, then, the mother, which refused to leave the side of its dying calf, could easily be captured. 'That shows the deep love a mother whale feels for her calf' (Marukawa 1941: 175). Then Marukawa goes on to relate the following tale, beginning with 'once I heard a story that goes like this' (Marukawa 1941: 175).

There was a young harpoon gunner, who first joined the crew of a catcher boat at the age of thirteen and later came to be praised for his skill. Eventually the harpooner married and became the father of a son. The harpooner was overjoyed and because all those around him congratulated him on the birth of his son he began to feel he was the happiest man in the world. In the meantime, the whaling season came around again and the gunner boarded his catcher boat feeling happier and healthier than before.

One day the catcher boat with the harpoon gunner on board, discovered two whales swimming far off in the distance. The boat made ready for the catch and the harpoon gunner stood by behind the gun mounted at the bow of the catcher boat. The boat approached the whales at full speed, but when they drew close enough to see them clearly, they realised that what they had thought to be two whales were actually a pair of whales with a suckling calf. The calf, shepherded between the two parent whales, was swimming as hard as it could and, to the harpoon gunner's eyes, the parents, perhaps because they had sensed the danger, seemed to be moving in even closer to protect the calf. In his surprise the harpooner forgot for a moment where he was, until he was returned to his senses by the angry voice of the captain. They had come right up to the whales.

The harpoon gunner, just as he had always done before, fired a harpoon at the calf and scored a direct hit. A cheer went up from the boat. The harpooner was beside himself with excitement, but he suddenly realised that the mother whale had come up to the side of the calf and seemed to tap it on the head with her pectoral fin and the next thing he knew was that the calf was covered with something white.

It was the mother's milk. The scene was just as if the mother were offering the calf her milk in a final gesture of love for her child. As he was looking at that whale and its calf, the image of his own beloved son and his wife floated up in his mind. Again, the piercing scream of the captain woke him from his reverie. The harpooner 'in his mind, once more bringing his hands together in prayer' (Marukawa 1941: 179) then shot dead the mother too.

Not long after that, the harpoon gunner threw in his job and took up a different occupation.

This is the story as Marukawa told it, but one wonders if this tale is based on a real incident he had seen or heard about somewhere. Or was it perhaps a piece of folklore spawned and narrated by workers engaged in the whaling industry. Of course, we have no way of knowing that. But, on the other hand, we can find scattered here and there in the pages of Marukawa's book, nationalistic narrative that accords with the current of the time. To take the section on the training of harpoon gunners, for example, Marukawa mentions the need to train fledgling harpoon gunners with 'military spirit' so they can fight against foreign countries as 'warriors of production,' to bring about whaling based on 'the Japanese spirit' in which 'the whole crew works with one heart' (Marukawa 1941: 128). The 'spirit' of 'warriors of production' is evident in the angry voice of the captain and in the cheer from the crew when the whale calf was hit. I think we can see this 'spirit' has the same content as 'the logic of overfishing.' It was this concept that the young harpoon gunner, struggling with his emotions at the sight of the mother whale and her calf, was unable to accept in his heart. Could it be that, in the contradictions represented in this storybook for children, Marukawa has unconsciously described his own complicated feelings towards whales?

Be that as it may, the guilt over killing whales, or the state of consciousness regarding the killing of whales, which falls outside the realm of rational argument, has clearly filtered through in the formation of the statements of those connected with whaling in Japan from the nineteenth century to the present day. I think it is important that, as long as we go on living as human beings, this kind of consciousness that can surface at any place and at any time, continues to live on inside us, often, refusing to be sublimated, finding expression in memorial services and the like. We should be continually moving to and fro' between the two positions of needing to kill in order to live and feeling guilt over killing. And while we may not be able to speak about this consciousness deep in our hearts, that is, I would venture to say, quite a different matter from not speaking, in order to improve one's own position or promote the interests of one's country.

6 The Politics of the Representation of 'Culture'

« Moi, j'affirme qu'ils ne savent ce qu'ils disent. »
Vous devez les voir comme ils sont, pas beaux, pas grands. « Pas
vrai », ajoute mon voisin.
Mais j'écoute attentivement, soigneusement. S'ils fredonnent
un chant, je fais attention de n'en point perdre l'air « comme on
perd ses billes, enfant ».
Vous qui lirez cela, défiez-vous de tout jugement.
Songez qu'ailleurs, tant d'hommes n'ont pas plus de sons que les
pièces de bronze dont on paie leur misère.

Paul Éluard

Overview of the argument

Plural relationships

In the discussion to this point we have considered the history of the
relationships between whales and human beings, focusing on Japan
in the nineteenth and twentieth centuries. Now, before proceeding
to the analysis of the discourse of anthropological research, which
is, as I indicated in the Prologue, the ultimate aim of this volume, I
would like to summarise what has come to light so far.

First, we can say that in Japan there has been a plurality of
relationships between human beings and whales (taking the order
cetacea as a whole). The commonly held view in regard to the
relationship between people and whales in Japan is that whaling is
a 'tradition' of 'the Japanese' that has been practised from the time
of the establishment of net whaling in the seventeenth century to the
present day. Careful examination of the historical record of these
relationships, however, revealed that this is not actually true. So,
let me try to bring together various examples of this relationship,
paying particular attention to the utilisation of the whale.

I can begin by giving the example of the use of whale meat or
whale oil from whales that have been killed or have died from other

causes. This can be active or passive. The active aspect is seen in whaling or in fishing with dolphins or porpoises. Net whaling was carried out during the Edo period in some areas such as northern Kyūshū, Kōchi and Wakayama. Norwegian-style whaling and factory-ship whaling were introduced later. And the taking of dolphins or porpoises, employing fishing techniques, such as driving them into inlets, which are thought to have existed since early times (before the Edo period), are still being practised in some areas today (Kasuya 1996) (Kasuya and Miyashita 1994) (See also. Chapter 4 of this book). In contrast, passive use implies the utilisation of the carcasses of drifting whales or whales stranded on reefs (See Chapter 4).

Next, there is the relationship that makes use of living whales. One example is the technique known as *sunameri ajiro* or fishing with finless porpoises, discussed in Chapter 3, which was formerly employed in the Inland Sea in the vicinity of Abajima Island in Hiroshima Prefecture. Another case of this kind of relationship can be seen in the rod and line fishing of bonito, when the presence of whales is used as an indication of the location of schools of bonito (See Chapter 4).

In addition to these, we might also include religious relationships, like those of some coastal fishing communities that see whales as manifestations of the god Ebisu and make them objects of pious devotion (See Chapter 2 etc.).[1] The origin of this belief lies in the fact that whales pursuing sardines or other fish which sustain local fishing communities drive them closer to shore making them easier for the fishermen to catch. We can probably see this as an example in which the fisherfolk benefit indirectly from (what they believe to be) the benevolence of whales, rather than the direct utilisation of whales we saw in the case of the finless porpoises. In addition, we should probably also include the lack of utilisation, as was the case in mountain communities in the days before the development of transport and preservation techniques.

These various relationships did not necessarily only occur in one place or at one time. Sometimes several occurred concurrently. For example, there were communities that both hunted whales and made use of live whales (Chapter 4, Notes). Also, although we did not cite any specific examples, it is assumed that some of those communities that considered whales objects of religious devotion would also have made use of stranded or drifting whales. Besides these relationships, we saw an example where the relationship probably differed according to the type of cetacean involved (Chapter 4). Further, there are bound to be individual differences from person to person and place to place in the way these relationships are played out or in the way whales are perceived.

As an appendix to these relationships, in addition to the connections to fishing, or what may be represented as 'folkloric phenomena' that were elucidated in these pages, there are those relationships in Japan today with the comparatively smaller species of cetacean raised in aquariums where they perform tricks for our amusement. And even in Japan we have seen the development of whale watching and dolphin therapy (Geisharen 1996; Miyoshi 1997). We must not overlook these trends, including those calling for the establishment of new relationships with whales.

As there have been various relationships of this kind at various levels, we can say that historically there has been, and there continues to be, a plurality of relationships between whales and human beings in Japan. For that reason, and I have said this before, the relationship between whales and 'the Japanese' cannot be restricted to whaling and whale meat consumption alone.

Incidentally, I think I need to explain why I use the term 'plural' (*fukusū*), instead of 'diverse' (*tayō*). Today we are coming to a consensus on respect for biological diversity. And at the same time, it seems to me, there are calls for respect for cultural diversity. But biological diversity is itself a diverse concept said to be comprised of three levels; gene diversity (diversity or genetic variation found within a single population or between different local populations), species diversity (the existence of many different types of species) and ecosystem diversity (the existence of a variety of ecosystems in response to differences in temperature, landform etc) (Higuchi 1996: 7–11). On the other hand, when we speak of cultural diversity, we do so on the premise that 'culture' exists among people who are regarded as constituting a group, whether that be the State or an ethnic group. Consequently, it is not appropriate to set up the idea of cultural diversity in analogy with biological diversity, as if one group like this were one species and the two could be compared side by side. Further, as we have seen in this book, there is not just one relationship between whales and 'the Japanese,' but many. So, I have used the term 'plural' in preference to 'diverse,' which I have avoided using because of its links with the concept of cultural diversity.

Let us now consider how these relationships between whales and human beings in Japan, which had taken so many different forms, developed in the process of the development of the Japanese whaling industry. But first let me summarise what we have discovered so far about the development of the Japanese whaling industry since the end of the nineteenth century.

Reduction to a single relationship

The establishment of the En'yō Hogei K.K. (En'yō Whaling Co.) and the Nagasaki Hogei K.K. (Nagasaki Whaling Co.) in 1897 sounded the opening salvo of Norwegian-style whaling in Japan. Then, we can say the Norwegian method became firmly established in Japan through the success of Nihon En'yō Gyogyō K. K. (Nihon En'yō Fisheries), the precursor to Tōyō Whaling, which was later to have a virtual monopoly of the Japanese whaling industry (See Chapter 1). We can identify two points that characterise the introduction and expansion of whaling at this time. One was the fact that the establishment of whaling stations in areas where net whaling had not been practised previously, resulted in friction and clashes with those already engaged in fishing in these areas. The most serious of these clashes was the riot in 1911 at the Same station of Tōyō Whaling, described in Chapter 2.

The other characteristic was the conjunction between the development of whaling and the contemporary currents of Japanese expansionism. The government of the day promoted the introduction of Norwegian-style whaling in the form of the Law for the Promotion of Pelagic Fisheries proclaimed in 1897 and this whaling method started in the waters around the Korean Peninsula, eventually to become a Japanese colony. Even after that, until the end of World War II, the whaling interests around the Korean Peninsula remained in the grasp of Japanese whaling companies (Chapters 1, 2 and 3).

In this way, the relationship between whales and people, had come to be dominated by whaling which had become a huge industry, deeply involved with the activities of the Japanese State and had penetrated, even to the point where it had caused friction and clashes with fishermen, regions, including colonies, where whaling, even on the scale of net whaling, had not previously been carried out. This activity led to overfishing and reduction in the number of whales. This is demonstrated, in particular, by the fact that the Asian populations of grey whales are now critically endangered, apparently as a result of the large catches of these creatures in the waters around the Korean Peninsula by the Japanese whaling industry (Chapter 3). Because of diminishing whale stocks, the Japanese whaling industry sought to establish other whaling stations, not only in the area of the Korean Peninsula, but also in eastern Hokkaidō, Chishima (the Kuriles) and Karafuto (southern Sakhalin Island) (See Chapter 2, Notes).

In 1934 Nihon Hogei K. K. (Nihon Whaling) was established when Tōyō Whaling was bought out by Nihon Sangyō K. K. (Nihon Production Co.). In the same year, this company became the first Japanese whaling company to carry out factory-ship whaling in the Antarctic Ocean. During this period in particular, we cannot overlook the influence of war. Whaling in the Antarctic was continued until the outbreak of war in the Pacific in 1941, as an important element of Japan's national strategy. The main focus at this time was on oil production, both as a means of securing foreign currency and as an important military resource, rather than on whale meat production, which was of secondary importance (See Chapter 1).

After Japan's defeat, GHQ gave its approval for the resumption of factory-ship whaling in the Antarctic in the 1946–47 season, as a solution to the country's food shortages. For this reason, and as a temporary phenomenon, there was a rise in the proportion of whale meat in total meat consumption (See Chapter 5). It was not until this period that whale meat became part of the everyday lives of 'the Japanese.' In the period prior to and during World War II, in spite of its promotion under the activities of the State, for strategic purposes and the waging of war, whale meat was by no means part of the daily diet across the entire country (Chapter 4). Behind the post-war spread of whale meat consumption, from those areas that had been involved in whaling, to cover the whole country, there was, in addition to the decline in coastal whaling, the factor of Japan's participation in the so-called 'whaling Olympics,' in which all the countries involved competed against each other to catch as many whales as possible within a fixed quota and which led to indiscriminate fishing of most of the larger species of whale in the Antarctic (Chapter 5).

The conclusion we can draw from the argument so far is that, in the history of whaling in Japan, the plural relationships that once existed between whales and the Japanese, have been reduced to the single relationship of whaling and eating whale meat, as a result of the massive growth in the whaling industry, due to its being made a 'national policy' from the end of the nineteenth century. Moreover, after Japan's defeat, when eating whale meat had become more or less an everyday occurrence, so too did this new single relationship with whales become the norm for 'the Japanese.' Further, we might say that in part the reduction to a single relationship had developed in lockstep with the process of the people becoming 'kokumin' (loyal subjects) for strategic purposes and the waging of war. But this simplification of relationships resulted in overfishing, particularly of the larger species of whales, which ultimately became a noose around the neck for the whaling industry, forcing it into decline from the beginning of the 1970s.

Given these historical developments, could we not perhaps treat Japanese whaling as indigenous whaling, even though nowadays this term implies small-scale whaling, which had became a large powerful industry in the process of colonial rule? To answer this question, let us turn to our ultimate goal, i.e. to critically investigate the discourse that sees whaling as 'culture' and Japanese whaling as equivalent to indigenous whaling. I should add that in the discourse analysis we are about to attempt here, we shall be focusing on the politics of representation as discussed in Chapter 5.

A critique of the 'whaling culture argument'

The 'whaling culture argument' of M. M. R. Freeman et al.

As I mentioned in the Prologue, when faced with the moratorium on whaling in 1987, the Japanese government persistently sought to continue whaling under the rubrics of scientific whaling and aboriginal subsistence whaling. The Japanese government submitted its plan for scientific whaling to the IWC in April 1987 and at the same time put forward the claim that the small-type coastal whaling currently practised in the four locations of, Abashiri in Hokkaidō, Ayukawa in Miyagi, Wadaura in Chiba and Taiji in Wakayama can be considered aboriginal subsistence whaling. What I have dubbed the 'whaling culture argument' was then mobilised to provide the theoretical and empirical justification of this latter claim.

So let us now consider the 'whaling culture argument' in more concrete detail. The Japanese government submitted its *Report on small-type coastal whaling* (Freeman *et al.* 1988)[2] to the IWC in 1988. In their report M. M. R. Freeman and his collaborators demonstrate how, in the places where small-type coastal whaling is practised, there is a non-commercial, ritual distribution of whale meat, a developed and varied 'food culture' based on whale meat, religious beliefs concerning whales which are manifest in festivals and in Buddhist memorial services, and so on. On the strength of this evidence they claim that in these areas whaling is of 'social, cultural and economic significance' (Freeman *et al.* 1988:83). And they conclude that since small-type coastal whaling can be thought to combine characteristics of commercial whaling and those of aboriginal subsistence whaling, the IWC should recognise a new category of whaling to accommodate it.

Let us now consider what Freeman *et al.* mean by the term 'whaling culture,' starting with their definition of 'culture.'

> Anthropologists generally use the word 'culture' to mean 'shared knowledge' transmitted through a socialization process from one generation to the next (Freeman *et al.* 1988: 28).[3]

Then they go on to explain 'whaling culture.'

> A whaling culture, such as that described here, may be defined as the shared knowledge of whaling transmitted across generations. This shared knowledge consists of a number of different socio-cultural inputs: a common heritage and world view, an understanding of ecological (including spiritual) and technological relations between human beings and whales, special distribution processes, and a food culture.
>
> The common heritage found in Japan's whaling culture is based on a long historical tradition. In this respect, it is primarily focussed on time, in that it relates myths, folk tales, legends and other narrative events concerning whales and whaling. (Freeman 1988: 75).[4]

And, the banning of commercial whaling, they say, places those communities with a 'whaling culture' in danger of collapsing. Consequently, they claim,

> [The residents of whaling communities] have great difficulty in understanding the imposition of a total moratorium which prohibits the catching of what seem to them to be abundant stocks of minke whale in Japanese waters, while at the same time having permitted the hunting of the endangered bowhead whale in North America. (...)
>
> There is a sense of bitterness about the present situation which residents in the whaling communities see as being a sanction specifically aimed at the Japanese by Americans. The criticism levelled at the Japanese for eating whale meat, for example, is interpreted as an assault on Japanese culture. The whaling issue has thus become a national symbol for the entire nation. (Freeman 1988: 78).[5]

However, we can point out quite a number of problems with these claims of a 'whaling culture.' As is clear from the previous quote, Freeman *et al.* emphasise history, by which they seem to mean continuity with the past. But that continuity exists only *à priori*. Even if we accept that such continuity can be demonstrated, we feel the need to ask by whom, for what purpose and in what manner was it formed. In other words, what were the dynamics behind its formation? Freeman *et al.* neither ask, nor provide the answers to, these questions. This we can elucidate from the following points. Firstly, in statements of Freeman *et al.* there are some distorted details regarding the history of whaling. Let us see how Freeman *et al* depict the unfolding of the history of whaling. They include maps of the 'Historical Developments in Japanese Coastal Whaling' in their report (Freeman 1988: 17).[6] According to these maps, Norwegian-style whaling started in present-day Yamaguchi prefecture and 'was diffused' through the 'movement' of whale catcher crews from there through western Kyūshū to Hokkaidō, then from Kōchi, through Wakayama and Mie prefectures to Chiba and eventually to the Miyagi and Iwate area. The major problem with this explanation is that it makes no mention of whaling in Japan's former colonies, particularly around the Korean Peninsula. As I have mentioned several times before, we can consider that the rebirth of Japanese whaling after the demise of net whaling occurred on the Korean Peninsula. Moreover, the references to 'movement' and 'diffusion' are altogether too mild to be accurate representations of the process, which would be better described as having been the result of colonial rule.

They also ignore the friction and clashes between whaling companies and local fisherman. Recalling the definition of 'culture' of Freeman *et al.* as transmitted 'shared knowledge,' we can say that the ideas and beliefs of the fisherman who opposed whaling also constituted a 'culture.' But the whaling companies took the belief of the fisherman of the time, that whales chased sardines into shore to make them easier to catch, branded it groundless superstition, and eventually supplanted these ideas with a different 'culture,' i.e. 'whaling culture.'

The reason Freeman *et al.* do not face up to the two points I have just mentioned, is probably because their research has the political objective of protecting Japanese whaling and therefore requires them to represent what they call 'whaling culture' as pure, innocent Japanese 'culture,' under threat from the unreasonable demands of 'the West.' The outcome of this approach is that in order to confront 'the West,' they represent the ideas of people in specific communities as those of 'the entire nation' of Japan (Freeman 1988: 78, the section quoted above). What arises from all this, alongside the problem of continuity, is the need to question the representation of wholeness, implied in the term 'the Japanese,' by representatives of only part of that whole. So, in order to find out, among other issues, whether this continuity and wholeness can be established as fact, let us now consider the work of Takahashi Jun'ichi, one of the central figures in the research of Freeman *et al.*

A critique of Takahashi Jun'ichi's 'whaling culture argument'
Takahashi Jun'ichi, in his 1987 paper (Takahashi 1987), adopts an instrumental approach that sees ethnicity as a political and symbolic means of informal group organisation. He analyses Taiji as an example of the social and political use of 'shared cultural traditions' within collectives that are not generally considered ethnic groups. In the paper, he elucidates how, every time external pressure is brought to bear on 'the Taijians' (*Taijijin*) through, for example, amalgamation with neighbouring towns and villages, the construction of a nuclear power plant or the prohibition on commercial whaling, symbols of identity (folk dance, local chronicles etc.) are created and mobilised to 'raise' the identity of Taiji as a 'whaling town' (*kujira no machi*). And he claims that this 'Taijian identity' has been put to good use as a positive force supporting village political strategy and activism to confront external pressure.

Takahashi strives to analyse objectively how the people of Taiji employ 'culture' politically and, further, as is obvious in the way he represents them as 'the Taijians' (*Taijijin*), he emphasises the unique position of Taiji within Japan, and particularly within those areas then engaged in whaling. But in the research of Freeman and his colleagues, the 'raising' of the identity of the people of Taiji is represented as follows.

> Whale memorial rites held by some social groups emphasize the unity of the community vis-à-vis outside forces perceived as hostile to its interests. The whale memorial rites held on April 29 for all of Taiji Town, sponsored by an association of former whalers (Taiji Hogei O.B. Kai) is a good example. It is of course no accident that these rites are held on April 29, the birthday of the reigning emperor. The timing is arranged to lend a patriotic significance to the rites. (…)
>
> On this occasion the memorial rites are not restricted to the religious actions of clergy. Speeches are made by important persons,

asserting their unflagging commitment to the continuation of
whaling and their absolute and unanimous opposition to the whaling
moratorium. (…)

In this way whale memorial rites call for closing ranks against
the outside in defense of a traditional mode of subsistence (Freeman
1988: 58–9 ([1989]: 147–8)).

Bearing in mind, of course, the different location and the different
circumstances, we can see here in the representation of the whale
memorial rites, points of analogy to the service in memory of the
Emperor Meiji, introduced in Chapter 2, in which speeches by
influential members of the community were given as an adjunct to a
religious ceremony. Comparing the two in this way, I think perhaps
what we can see here, in the above representation of Freeman *et al.*, is
the perfected form of the process of turning fishermen into '*kokumin*'
(loyal subjects) and linking the process to whaling, as we observed
in Chapter 2. And anthropologists like Freeman and Takahashi
have enthusiastically and positively represented this identity as that
of the '*kokumin*' (nationals, citizens) or 'the Japanese' (*Nihonjin*).
But, when we consider that the historical development of Japanese
whaling in the modern period, took its direction from Japanese
expansionism, it was, in a sense, this circumstance that led to the
formation of *kokumin* (nationals of the Japanese State) and *Nihonjin*
(the Japanese as an ethnic group). If Freeman and his collaborators
had taken this fact into account and been a little more prudent in
their representation, they may not have ended up with the kind of
nationalism we can see in the quoted passage. But their not doing so,
we can assume, is to develop their argument that 'whaling culture'
is Japanese *culture*, that is to say, that a phenomenon represented as
'culture' does not exist independently, but is one of the elements that
make up *culture as a whole*. And, as we shall see below, Takahashi's
own 'whaling culture' argument develops in the same way, rejecting
the individual and focusing on the whole (Takahashi 1991; 1992).

In his more recent papers, Takahashi begins with the following
manipulative definition of 'culture,'

The integrated system of consolidated knowledge, skills and social
organisation required for people to seek, find, obtain, prepare/
process, then distribute and consume, the resources of the ecological
environment in which they live (Takahashi 1992: 19).

He goes on to claim,

When we can recognise a phenomenon in which a particular group of people adopts a unique lifestyle based on whaling activity that is organically linked to various social, economic, technical and spiritual aspects of the group' we can call it 'whaling culture' (Takahashi 1992: 21).

He concludes that Japan has a unique 'whaling culture,' which should be maintained in the interests of 'cultural diversity' (Takahashi 1992: 161) and for the future of human beings as a whole.

Further, Takahashi argues along the following lines regarding the development of whaling in Japan. First, he points out the importance of a 'support system' in the modernisation of whaling and 'the diffusion of whaling culture' (Takahashi 1992: 82). He claims that the failure in the opening up of 'new fishing grounds' at the end of the nineteenth century was due to the fact that while many whale hunting specialists were sent off to the fishing grounds, an on-shore 'support system' for appropriate processing and efficient distribution and consumption of the whale carcass was not provided (Takahashi 1992: 75–6). Consequently, by the beginning of the twentieth century, the provision of an on-shore support system became an indispensable component in the exploitation of new fishing grounds. At first, the whaling companies brought in professional whalers and flensers from other whaling areas, but employed local labour for the other processing tasks, but eventually local people also came to be engaged in specialist whaling work (Takahashi 1992: 82–5).

Takahashi also makes the claim that both large type coastal whaling and factory-ship whaling are a continuation of 'the whaling culture' that goes back to net whaling. The basis of his claim for continuity with net whaling is that 'most of these common features are found in those fundamental areas that characterise Japanese whaling' (Takahashi 1992: 113). These continuities can be found, he says, first, in the fact that a clear division, evident in differences in knowledge, techniques, and the affiliation with local and kinship groups, is maintained between whale-catching activity and processing activity. As further support for his claim, Takahashi cites the continuity and conservatism of the technology and techniques employed in processing based on the stability of the 'tradition' of whale meat consumption, and the reciprocity between the operators (whaling groups or whaling companies) and the areas where the fishing grounds and processing facilities are located (Takahashi 1992: 90–116).[7]

In other words, Takahashi sees the reduction to a single relation-ships in a positive light, in the form of 'the diffusion of whaling culture.' Further, by not confining himself to small-type coastal whaling, but also including large-type coastal whaling and factory-ship whaling in Japanese 'whaling culture,' he seeks to justify the continuation of whaling on the grounds of 'cultural diversity.' The problem with this argument is, first of all, in the same way as Freeman *et al.*, Takahashi's use of the gentle term 'diffusion' to represent the implantation of a new 'culture.' Did he think there was no domination or friction in the introduction of whaling to areas, like those on the Korean Peninsula and elsewhere, that had not experienced whaling on the scale of the net whaling groups? Here let us look again at the actual composition of the workers involved in catching and processing activities, as I did to a certain extent in my criticism of the 'whaling culture argument' in Chapter 1. In general, whaling under the net whaling groups was closely bound up with the local community, and, in keeping with the social structure of the time, it was laid out on class lines based on status and heredity. But the workers in the early modern Japanese whaling industry were made up of a hierarchy in which Norwegians were in the top rank as the purveyors of the technology and skills required, not only for catching whales, but also for processing them. Next came the Japanese, who, as managers gradually came to replace the Norwegians as the suppliers of new technology, and finally the Koreans, the majority of whom remained at the bottom of the hierarchy. In short, we can see how a new social structure based on nationality was formed, as whaling became a major industry with the start of Norwegian-style whaling off the Korean Peninsula. Consequently, the very formation of a modern whaling '(support) system' different from that of the net whaling groups, can be described as a process of colonial domination, which included, as I have described in Chapter 2, friction and confrontation. And the 'diffusion of whaling culture,' deserves to be criticised as a violent process of reduction of relationships with whales, which led to indiscriminate overfishing.

Further, I feel I have already included in these pages something of a refutation of Takahashi's argument that there is a continuity in 'whaling culture' from net whaling, to large-type coastal whaling and factory-ship whaling. For example, in regard to the maintenance of a clear division between catching activities and processing activities, I gave an example in Chapter 1 of how, under net whaling, those who had carried out the catch were later assigned to processing activity. Besides, in regard to the continuity and conservatism of processing activities arising from the stability of 'the tradition' of whale meat consumption, Takahashi overlooks the fact that this 'tradition' was constructed. In addition, I showed how processing activity under Norwegian-style whaling was not simply a continuation of the processing carried out in 'the barn' (*nayaba*) of the whaling groups (Chapters 1 and 4). Although Takahashi cites the reciprocity between the operators and the areas in which the fishing grounds and processing facilities are located, he does not give appropriate recognition to the fact that often these relationships are forged, either in order to set up whaling in a new area, or to ease tensions that have arisen in an area. I also pointed out that, in as much as the operators are commercial enterprises, when they can no longer catch whales in an area, they close down their stations and move on to set up elsewhere (Chapter 2).

Apart from these criticisms, however, there is a more fundamental problem with Takahashi's analysis. How, for example, are we to interpret the following? In the 1970s there was a problem with a 'pirate whaling ship' named the *Sierra*, a combined whale catcher and factory ship, which hunted banned blue whales and humpback whales in the Atlantic Ocean and exported their meat to Japan. In 1978, the crew of the *Sierra*, involved in the illegal trade, comprised a Norwegian captain and harpoon gunner and the rest of the crew were South Africans. In addition, there were four Japanese whale meat inspectors on board who had formerly been engaged in processing for Japanese whaling companies. In this way, it is said, Japanese whaling companies had been closely connected to the *Sierra*, which carried on its illegal whaling for approximately ten years, and regularly had from four to six Japanese whale meat inspectors on board to supervise the processing of whale meat (Hara 1993: 63–76). If we are to follow Takahashi's argument, the *Sierra*, which was producing whale meat 'designed for the Japanese market,' should also be seen as continuing the 'whaling culture' that began with the net whaling groups. Or perhaps, since it was Norwegian-style whaling with a Norwegian harpoon gunner, it was a continuation of Norwegian 'whaling culture.' Or could it be, 'given that there were not only Japanese involved, that it did not belong in Japanese 'whaling culture' going back to the days of net whaling' at all? However, as I mentioned before, Norwegian-style and factory-ship whaling in modern Japan were by no means conducted by Japanese alone. And, as I made clear in Chapter 1, the composition of the workers involved in catching and processing activities at the time of the introduction of *Norwegian-style whaling*, was not so very different from that of the crew of the *Sierra*.

The problems with Takahashi's argument stem from his methodology. From among the various changes that occurred in the history of whaling, he seeks out and abstracts what he sees as the unchanging 'elements that can be considered folkloric or anthropological,' that form the basis of both large-type coastal whaling and factory-ship whaling. His mistake is that he then makes the substantialist assumption that these are a continuation of the 'whaling culture' of the net whaling groups. A much more plausible interpretation, I would have thought, is that large-type coastal whaling and factory-ship whaling, arising as they did from the changes in organisational structure and introduction of new technology mentioned above, should be regarded as completely different in their formation from net whaling. If I may be permitted to comment further, I would also like to point out the danger, in designating any large industry the 'culture' of a particular nation state or ethnic group. Even though it may not extend to the point of piracy, there is the fear that it will be giving *carte blanche* for unbridled economic activity of that state or ethnic group.

It should be clear from what I have said here that I think there is a specific political motivation in the development of Takahashi's research. It is to protect all levels of whaling practised in Japan and to advocate this view domestically and abroad. Moreover, Takahashi substantialises the state and the ethnic group, and represents the reduction to a singular relationship as 'culture.' Consequently, his call for 'cultural diversity,' ends up becoming, as I have just pointed out, a demand for 'cultural diversity based on the unit of the state and ethnic group.' But, in reality, there is a plurality of relationships between 'the Japanese' and whales. So, in this sense, we cannot make genralisations about 'Japan' or 'the Japanese.' Consequently, if indeed Takahashi's call for diversity does include all the various kinds of relationships between whales and human beings, then it contradicts his claims regarding the 'diffusion' or 'transplantation' of 'culture.'

Facing up to plurality

Ultimately, we can say that weaknesses arise, in both the research of Freeman and his colleagues and in Takahashi's own work, because of their failure to gain an accurate grasp of the development of whaling in early modern and modern Japan. In short, in as much as their research has the political objective of protecting the Japanese whaling industry, its development must not be coercive; in the same way that European colonial rule over indigenous peoples was coercive. And in *Japan* as a whole, relationships between whales and human beings need to be unified into whaling. Further, there can be no construction of a whaling relationship, nor any gap between present-day whaling and that practised in the past. For these reasons, Freeman and his associates, while on the one hand 'not telling' about the facts of the development of modern Japan or modern Japanese whaling, on the other hand perform a leap of logic which takes a representation of just that one part of the industry known as 'small-type coastal whaling,' and proceed to apply it to *Japan* as a whole. What is more, Takahashi arbitrarily establishes 'fundamental' elements to demonstrate continuity through the whole of Japanese whaling.

Alternatively, we can think of these difficulties as arising from the fact that both Freeman *et al.* and Takahashi define 'culture' as having wholeness and continuity, as indicated by Clifford introduced in Chapter 1 – though Freeman *et al.* stress continuity, whereas Takahashi places the emphasis on wholeness. In short, the development of whaling in modern Japan does not fit this definition of 'culture,' nor can it be adequately explained with this model. What then if they had defined 'culture' in terms of so-called hybridity, not an organic whole or independent body at all, but a number of elements gathered or tied together (Clifford 2003: 504–6)? Would that have solved the problems in their analyses? Arguably, the development of whaling in modern Japan could be regarded as a hybrid phenomenon, given the range of people and techniques involved in the process and the way it transformed a plurality of relationships into a single relationship. But the representation of hybridity, when 'culture' and 'tradition' are changed through colonisation or globalisation, is used not only to give a positive slant to the strategies that subjugated people, indigenous peoples for example, appropriate from their oppressors. We need to remember that hybridity can also be used to justify the process in which the rulers appropriate all manner of things from others and make them their own.[8] Anyway, even if we do regard Japanese whaling as a hybrid, the questions, what forces underlie its formation, and, how is each individual pressed into becoming part of the conglomerate we call 'culture' or 'tradition,' remain unanswered. Also, the problem of the political context of its representation is left unsolved.

We would expect a number of counter arguments to be put forward in response to my criticism of the 'whaling culture' argument. One might be that, even if we accept that whaling and the consumption of whale meat are relatively recent phenomena, wouldn't we still be justified in regarding the shared experience of 'Japanese citizens,' over a specific period of time as 'culture?' In regard to this, I could probably reply that knowing that 'culture' and 'tradition' are constructed enables us to say we have the ability to create certain relationships – not, of course, how many years must pass before a relationship becomes 'culture' or 'tradition.'

And, even if a particular group shared an experience over a specific period of time, that is not to say that it was necessarily a good thing (for example, in harmony with environmental protection or the like). As I mentioned before, after defeat in Word War II, the simplified single focus on whaling and whale meat consumption became a fact of everyday life for 'the Japanese.' This resulted in overfishing, particularly of the larger species of whale. It follows, therefore, that we need to create a new relationship with whales.

Another objection might question the way I treat what has today come to be called 'identity politics,' like the activity of 'the Taijians,' alluded to by Takahashi, in which a minority group or a locality invokes 'culture' and 'tradition' to confront the majority or the State ruled by the majority. As we have just seen, some anthropologists have enthusiastically and positively represented the Taiji people's identity as that of 'the nation' or 'the Japanese.' But, *contrary to the representation of Freeman, Takahashi and others*, the Taiji people may well have established their identity as a frame that permits them to make strategic changes according to the circumstances, so that at certain times they are 'the Taijians' and at other times they are 'the Japanese.' And researchers in the camp of strategic essentialism may well represent identity as this kind of strategically changeable frame, regardless of the realities of the phenomena involved. But it can lead to problems if no single individual is permitted to deviate from the frame of identity, or if any conflict in the mind of an individual, including those that waver between identities, is interpreted as a calculated 'use of a weapon.' For this reason, in Chapter 2, I considered ('internal') framing analogous to 'collective identity' and throughout this book I have considered the individual and individual conflicts. In other words, the plurality of relationships I have elucidated here, is not merely a matter of taking each area as a separate case, but covers the various relationships of each individual living in each area, including complicated feelings about whales. Recognising the plurality of these relationships is, for the individual, a way to live with these conflicts, and for groups, from the level of the local district to the international arena, as people living together, it is important that those whose relationship takes the form of whaling, recognise the existence of those who do not have (nor wish to have) such a relationship, and at the same time, for those who do not have (nor wish to have) a whaling relationship, to recognise the existence of those who do. Perhaps this point needs to be made more clearly. I am, apparently, both legally and socially, regarded as Japanese. Yet, I feel absolutely no need for whaling. On the other hand, as long as there are people who think that whaling is important in their everyday lives, I must face up to them with sincerity.

Conclusion

Above, I have clarified the history of the relationships between whales and human beings in early modern and modern Japan and criticised the 'whaling culture argument.' Now, in the light of this history, I should like to bring this book to a close by speculating on the future relationships between whales and human beings.

There are several political debates currently under way in Japan regarding the future of our relationships with whales. On the basis of the facts I have clarified in these pages, together with the points argued above, the following fundamental posture needs to be taken *vis-à-vis* these debates.

1. The protection of wildlife, in the sense of preserving the environment and not in the sense of maintaining 'a resource.'
2. Maintenance of a plurality of relationships between whales and human beings.

This would mean, considering whaling today, that only extremely limited catches would be permitted. There would need to be a complete overturning of the thinking that has prevailed to this point. That is, rather than thinking, quotas or no quotas, how can we catch as many whales as possible, we should be asking, from the outset, and on the premise that we can take only a very small number of whales, which is the most suitable body to carry out this task. If this could be achieved, the most practical outcome would probably be to follow the advice of Kitō Shūichi and Obara Hideo and recognise small-type coastal whaling or the pursuit of minke whales within territorial waters, but to cease whaling in the Antarctic Ocean (Kitō 1996: 164–6; Obara 1996: 123–37). Given the *status quo*, this would mean recognising the four areas mentioned previously as sites of small-type coastal whaling. My reason for advocating this course of action is not because the whaling of those areas was Japanese 'culture' or 'tradition.' Actually, whaling was not introduced into Abashiri and Ayukawa until the twentieth century, after the advent of Norwegian-style whaling (Freeman 1988: 28)[9] (See also Chapter 2 of this book). I simply feel that a plurality of relationships should be maintained, together with protection of wildlife and as long as it does not cause damage to the environment.

On the other hand, those with connections to the whaling industry advocate a continuation of Antarctic whaling, along the lines of the old Monopoly Corporation under national government control and administration, with the profits generated being applied to marine mammal research and the protection of the Southern Ocean environment (Kawabata 1995: 238–53; Nagasaki 1990b). But I feel compelled to reject this proposal, not merely because I doubt whether it would be acceptable to people living in Japan, including those opposed to whaling, but more so, because it is an extension of the whaling regime under Japanese expansionism in the first half of the twentieth century. It is a case of the State, once again, restricting the relationship between whales and 'the Japanese' to one of whaling and whale meat consumption. Further, R. L. Friedheim predicts that, in the future, only the Japanese will be whaling in the Antarctic Ocean. Given this fact, he says, it is important, when they are participating in negotiations, for representatives of nations and 'cultures' who do not give priority to the seas and oceans, to correct their misunderstanding of Japan's maritime activity. As an island nation Japan is dependent on the sea and is constantly worried about securing its food supply. He proposes that a fund, administered by the IWC, be established to finance various activities concerning whales, though he does not specify which organisation should carry out the task, and that Japan be permitted, as an exception, to conduct whaling within the Antarctic Ocean whale sanctuary, on the condition that it paid a whaling fee into the fund (Friedheim 2001: 320–1, 327–31). But Friedheim's 'understanding of Japan' is extremely shallow and stereotypical. And he ignores the plurality of relationships. For this reason his proposal has the same problems as that of the whaling industry mentioned above.

Moreover, even if (small-type) coastal whaling is to be continued, this should only be done after there has been a clear expression of regret over the expansionist development of Japanese whaling and indiscriminate overfishing. And, as Kitō says, and in line with the suggestion made when small-type coastal whaling was characterised as 'community-based whaling' in the IWC (Fujishima and Matsuda 1998: 120–1), to prevent overfishing, it is probably necessary to introduce regulations regarding distribution, such as restricting the amount of whale meat and other whale products permitted to be sold outside the (small-type) coastal whaling areas. Similarly, whale products should not be imported from foreign sources. To ensure that this policy is followed, the present thinking about the relaxing of the restrictions regarding whales in the so-called Washington Convention should be reversed to allow the rules to be tightened. Within this kind of context, it would be possible to support Ireland's proposal to the 1997 general meeting of the IWC, namely that whaling in the non-territorial waters of the open sea be banned. Further, even if a ban were placed on whaling in the open sea, the IWC should probably continue to exist as the body for international dialogue and whaling regulation. To carry the argument further, the actions mentioned above can be considered as part of the trend towards the creation of an ideal fishing regime as a whole, in which coastal fishing is revitalised and economic strength does not result in the plundering of marine products around the world.

In contrast to this approach, H. N. Scheiber, while taking an anti-whaling position, is prepared to recognise, as an exception, the 'cultural' demands of whaling of indigenous people, who have been victimised historically. He reaches the conclusion, however, that the Japanese towns that engage in coastal whaling should not be regarded in the same way as these indigenous communities. Scheiber argues that they should not be permitted to continue whaling on the grounds that, like other areas of Japan, they profited from Japanese military expansion and exploitation of colonial territories; are today reaping the benefits of Japanese economic growth, and are inhabited by people ethnically no different from other Japanese (Scheiber 1998). Scheiber's argument, emphasising the historical background and development, is in many respects similar to my own approach. That is not to say, however, that he has correctly appraised 'culture' and 'tradition' *per se* from a historical point of view, either in abstract or concrete terms. Neither does he consider the plurality of relationships. His opposition to Japanese whaling seems to be based on a single essentialist understanding about what constitutes 'Japan.' In this regard, and although he reaches entirely the opposite conclusion, I think we can say he is standing on the same ground as Friedheim.[10]

Further, D. G. Victor, while supporting the whaling moratorium on the one hand, recognises specific aboriginal subsistence whaling on the other. Moreover, he sees the IWC *status quo*, with Norway pursuing commercial whaling and Japan engaged in scientific whaling, as 'the Pareto optimal solution' (the stage at which one party's gain can only be made at the expense of another party's loss). For this reason Victor argues the current whaling supervisory regime requires no change, other than an amendment to the policy of indigenous rights, which would include allowing Japan's coastal whaling communities to be treated fairly in the same way as those engaged in aboriginal subsistence whaling (Victor 2001). One point in Victor's case that calls for further attention concerns the hegemony of the United States. Victor, from the standpoint of realism in international politics, takes a positive view of American domination which props up the current 'Pareto optimal solution' with the threat of sanctions. Certainly, it may be unrealistic to simply recognise the plurality of relationships and issue a 'policy statement' that ultimately appeals to people's tolerance. But, part of the reality is that hegemony, particularly the use of violence, has, historically and up to the present day, silenced minorities and devastated the environment. Even now, some continue to wield this kind of power and, no matter how much they mouth the words 'multiculturalism' or 'cultural diversity,' they are not facing up to this plurality, but are just repeating, in another form, the history we have analysed critically in this book. In short, creating a *new relationship with whales* also means that we go on hoping not to depend on this kind of power.

Of course, this does not mean that Japan should replace the United States of America as the dominant force in whaling or the fisheries industry. But assertions of 'culture' and 'tradition' can very easily be equated with nationalistic aspirations for hegemony.

Postscript

This book is an expanded and amended version of my doctoral thesis, *Kingendai Nihon ni okeru kujira to ningen no kakawari ni kansuru rekishi-shakaigakuteki kenkyū* (A historical sociology of the relationships between whales and human beings in early-modern and modern Japan) presented to Kyoto University in 2002. Both the thesis and this book have been fundamentally rewritten since, but the following chapters are based on previously published papers, as follows.

Prologue and Chapter 6, 'Hogei mondai ni okeru "bunka" hyōshō no seijisei ni tsuite' (On the politics of the representation of 'culture' in the whaling issue), *Kankyō shakaigaku kenkyū* (Journal of Environmental Sociology), 1998, 4: 219–34.

Chapter 1, 'Kindai Nihon hogeigyō ni okeru gijutsu dōnyū to rōdōsya' (The introduction of technology and the workers in the whaling industry in early modern Japan), *Kagakushi kenkyū* (Journal of the History of Science, Japan), 1998, 205: 1–16.

Chapter 3, 'Sangyō, hogo, tennen kinenbutsu – kujira rui no shitei o megutte' (On the designation of whale species as natural monuments – how far does industrial use come into consideration?), *Seibutsugakushi Kenkyū* (The Japanese Journal of the History of Biology), 2000, 65: 33–46.

Chapter 4, 'Kindai Nihon ni okeru geinikushoku no fukyū katei ni tsuite' (The promotion of whale meat in early modern Japan), *Kaihatsu to Kankyō* – Ōsaka Gaikokugo Daigaku Gaikokugo Gakubu Kokusai Bunka Gakka Kaihatsu-Kankyō Kōza (*Development and the Environment* – Division of Development and Environment Studies, Department of International Studies, Faculty of Foreign Studies, Osaka University of Foreign Studies), 2001, 2: 1–18.

It goes without saying that I have put all my knowledge and effort into producing this book. Even so, for a variety of reasons, I am sure there remain sections where the argument is deficient or I have failed to make myself clear. I would be very grateful to receive critical comments from my readers in this regard.

I would now like to take this opportunity to express my gratitude to the many people without whose support my research would never have come together into a book.

My academic supervisor, Professor Soda Osamu was a great help to me, not only with the production of my doctoral thesis, but throughout the period of my graduate studies.

Thanks to the extremely high degree of freedom Professor Soda maintained in his Laboratory of the Philosophy of Agricultural Science in Kyoto University, even I, from outside the immediate discipline, was able to pursue my own research interests to my heart's content.

I received very valuable comments from Professors Niiyama Yōko and Noda Kimio, who gave up precious time in their busy schedules to examine my dissertation. In addition, my current mentor, Professor Suehara Tatsuro, and Mr Ōishi Kazuo, Assistant Professor in the Laboratory of the Philosophy of Agricultural Science, not only gave me copious advice in seminars and elsewhere, but also encouraged me in various ways when I was at a loose end after gaining my doctorate.

When I was a postgraduate student I benefited greatly from the wise instruction of Mr Akitsu Motoki and Mr Sakiyama Masaki, both of the Laboratory of the Philosophy of Agricultural Science. They not only taught me a lot about research itself, but also how to live the life of a researcher. It strikes me as something of a miracle that these two individuals, so different in research styles and modes of living, should have come together in the same Laboratory. And that I, periods of fear and trembling not withstanding, was able to pass such happy days with them.

Further, although they are too numerous to name individually, I would like not only to acknowledge here in writing the constant stimulus I received from all those people I encountered in the course of my research, particularly my friends and seniors in the Laboratory of the Philosophy of Agricultural Science, but also apologise for all the inconvenience I undoubtedly caused them. It is not in my disposition to speak frankly about myself, so there may be some among my colleagues who to this day wonder what I am thinking about. Nevertheless, in the sordid confines of the seminar room, they certainly helped me out on numerous occasions.

Finally, I would like once again to thank Mr Shimoda Katsushi, the President of Tōshindō and Mr Mukai Tomo'o, who took on the task of editing, for agreeing to publish my book despite the fiercely competitive situation of the publishing world.

Hiroyuki Watanabe

Notes

Prologue

1 Translator's note: Also known variously as the Antarctic Ocean, the Great Southern Ocean and the South Polar Ocean. In the current whaling debate, Southern Ocean seems to be the most commonly used term, whereas Antarctic Ocean is more often used to describe the period up to the 1987 moratorium on whaling. I have translated the Japanese Nankyokkai as Southern Ocean in the context of contemporary debates, but Antarctic Ocean for earlier periods.

2 Japanese translation, Maehata Masayoshi et al., [1996]: 36–7, 68–71.

3 The exact number of species in the order of whales is not definitely known as there are differences of opinion regarding classification and there may yet be undiscovered species, but at present there are around eighty known species. As the International Convention for the Regulation of Whaling does not define the term 'whale,' some of the governments of member countries claim that all species falling within the order of whales come under the jurisdiction of the IWC ('Taxonomy of Whales': *http://www.iwcoffice.org/ conservation/ cetacea.htm*).

In contrast to this position, the Japanese government recognises the IWC's control over only the following thirteen species: the North Pacific right whale, the bowhead whale (both in the family balaenidae); the pygmy right whale (Family neobalaenidae); the grey whale (Family eschtichtidae); and six species in the the blue whale family (balaenopteridae) (in order of size, the blue whale, the fin whale, the humpback whale, the sei whale, Bryde's whale and the minke whale); the sperm whale; and the northern bottlenose and southern bottlenose whales in the beaked whale family (ziphidae). For this reason, the Japanese government considers that Baird's beaked whale of the family ziphidae and the pilot whale of the family delphinidae, both taken today in Japanese coastal waters, fall outside the jurisdiction of the IWC, and imposes its own catch restrictions on these species (Fujishima 1998: 115–16; 'Geirui no hokaku nado o meguru naigai no jōsei heisei jūgonen shichigatsu' (Internal and external conditions surrounding the taking of whales etc, July 2003), *http://www.jfa. maff. go.jp/ whale/document/brief_explanation_of_whaling_jp.htm*; 'Kogata hogeigyō ni kansuru kiso chishiki' (Basic facts about the small type whaling industry), *http:// homepage2.nifty.com/ jstwa/kisochisiki.htm*. Moreover, those species the Japanese government recognises as coming under IWC control are, for the most part, those comparatively larger whales that are important to the commercial whaling industry. Incidentally, the IWC divides the right whales into three species, the minke whales into two and calls the bottlenose whale the 'northern bottlenose whale' 'Taxonomy of Whales' *http://www.iwcoffice.org/conservation/cetacea.htm.*

4 According to the Japanese government's classification, at the time of the implementation of the moratorium on commercial whaling in 1987, and on the basis of the types of whales caught, the scale of the ships used and the methods of processing the carcass (whether on a factory ship or a facility on land), Japanese whaling was divided into the three types, large scale coastal whaling, factory-ship whaling and small-type coastal whaling (Freeman 1988 ([1989]: 23–5)). Of these, small-type coastal whaling used small boats in coastal waters to catch relatively smaller whales like the minke, Baird's beaked whale, pilot and killer whales.

5 In addition to commercial whaling the IWC also recognises scientific whaling and aboriginal subsistence whaling. Notice that the IWC specifically appends the word 'aboriginal' to subsistence whaling.

6 In accordance with its research findings submitted in 1998, the Japanese government is said to have requested a minke whale quota for its small-type coastal whaling industry along the following lines. During the period 1986–7 it claimed that its small-type coastal whaling fell under the category of aboriginal subsistence whaling and in the period from 1988 to 1992 it asserted that it came under a new category separate from both commercial whaling and aboriginal subsistence whaling. So, since 1993, it has held that small-type coastal whaling is community-based whaling and sought as far as possible to downplay its commercial aspect (Fujishima and Matsuda 1998: 120–1).

From the time of the 55th annual meeting of the IWC in 2003, the Japanese government, as its request for a 'temporary relief quota' for small-type coastal whaling had been voted down for fifteen years, has been requesting a commercial whaling quota of 150 minke whales for small-type coastal whaling and 150 Bryde's whales for large-type coastal whaling, in order to contribute to local communities and work towards the goal of achieving an RMS (Revised Management Scheme) ('Dai 54kai Kokusai Hogei Iinkai (IWC) nenji kaigō kekka' (Results of the 54th annual meeting of the IWC), *http://www.jfa. maff.go.jp/ release/14.05.31.7.html*; 'Dai 55kai Kokusai Hogei Iinkai (IWC) nenji kaigō kekka' (Results of the 55th annual meeting of the IWC), *http:// www. jfa.maff.go.jp/ release/15.07.07.1.html*; 'Dai 56kai Kokusai Hogei Iinkai (IWC) nenji kaigō honkaigō no kekka ni tsuite' 2004). However, at the 57th meeting in 2005, though the Japanese government planned to again request yearly quotas of 150 minke and 150 Bryde's whales, as it had done the time before and the time before that, it actually only asked for a commercial quota of 150 minke whales and made no mention of a quota for Bryde's whales ('Dai 57kai Kokusai Hogei Iinkai (IWC) nenji kaigō sōkai no kaisai ni tsuite' (On the opening of the general assembly of the 57th annual meeting of the IWC), *http://www.jfa.maff.go.jp/ release/17/050617IWCannualstart.pdf*; 'Dai 57kai Kokusai Hogei Iinkai (IWC) nenji kaigō sōkai no kekka ni tsuite' (On the results of the general assembly of the 57th annual meeting of the IWC), *http://www.jfa.maff. go.jp/ release /17/17.0624.02htm*; '2005 Meeting' *http://www.iwcoffice. org/meetings/ meeting2005.htm*).

Nevertheless, the Japanese government is now claiming that 'small-type coastal whaling in our country has important socio-economic and historical significance to regional communities, in the same way as aboriginal subsistence whaling does. This is clear from the analyses carried out over the past ten years or more by many foreign cultural anthropologists and in over forty academic papers published on the subject. With the protection of this tradition in mind, Japan, since 1988, has been requesting a temporary whaling quota of 50 minke whales a year for its coastal industry' ('Hogeihan no kihonteki na kangaekata' (The fundamental thinking of the whaling section), *http://www.jfa.maff.go.jp/whale/assertion/assertionjp.htm*). Further, the RMS (Revised Management Scheme) is a combination of the RMP (Revised Management Procedure), a method of determining safe whaling quotas from the confirmed 'size of the resource,' and a system of audit and surveillance to ensure that the agreed catch quotas are not exceeded. ('Revised Management Scheme': *http://www.iwcoffice.org/conservation/rms.htm*).

7 In this book, my use of the expression 'relationship' (*kakawari*) for the relatedness of human beings to living creatures (or Nature) follows Kitō Shūichi's precedent (Kitō 1996: in particular 120–31). I myself use the term '*kakawari*' not as an essentialised entity, but in the sense of a kind of on-going process continually under negotiation. In my use of the term I want it to convey the connotation that the relationship between human beings and other living creatures is different again from the relationship one human being has with another human being.

8 Taken from Perry Anderson's paper included in Friedlander ed. 1992, Abridged Japanese translation, Tadao Uemura et al. [1994], *Aushuvittsu to hyōshō no genkai*, Miraisha, pp. 136–7.

Chapter 1

1 This way of thinking is more obvious in Morita Katsuaki, who unreservedly cites Takahashi's research.

2 An accident that occurred in 1878 when 124 fishermen perished after setting sail in bad weather in pursuit of a right whale with a calf. (Kumano Taijiura hogeishi hensan iinkai 1969: 341–9; Morita 1994: 320–4).

3 There is also a description to the effect that operations commenced in December 1898 (Mishima 1899: 26), but I have opted for 1897 when whaling permission was sought and received (Torisu 1999: 336, 353–4; Tōyō Hogei K.K. 1910: 189–90).

4 Tōyō Hogei K.K. claims that the Hogei Kabushiki Kaisha of Nagasaki commenced Norwegian-style whaling in 1898 (Tōyō Hogei K.K. 1910: 228–9), but given that the name of the founder of this Hogei Kabushiki Kaisha and the name of the company's ship, (Hatsutaka Maru), match, we can assume that this is the Nagasaki Hogei K.K. described by Torisu Kyōichi. Further, we know from primary source materials that the application and permission to commence whaling date from 1897. Consequently, I have adopted the 1897 date.

5 After this Enyō Hogei introduced two sailing ships described as 'flensing ship cum freighter' (Chōsen Gyogyō Kyōkai 1900: 16).

6 Tosa's net whaling was carried out by the Ukitsu and Tsuro whaling groups. Both communities are situated on the Muroto peninsula. These two groups operated, one at a time, in the fishing grounds off the Muroto peninsula and off Kubotsu on Cape Ashizuri, alternating every other year (Izukawa 1943: passim, 1973c: passim).

7 In 1799, even within the provinces of Hizen, Tsushima and Iki alone, there were seven whaling guilds and eighteen fishing grounds (Hidemura 1952a: 58–9).

8 For accounts of Count Keizerling Pacific Whaling Company and the Holme Ringer Company, see Park Koo-Byong and Tōyō Hogei K.K. ed. (Park 1995: 181–21; Tōyō Hogei K.K. 1910: 185–8, 191–2).

9 The reason the earlier En'yō Hogei and Holme Ringer Company were unable to continue whaling can be attributed partly to the fact that they were unable to obtain concessions from the Korean government (Chōsen Gyogyō Kyōkai 1900: 16).

10 As with the Daiichi Chōshū Maru, En'yō Hogei's Hōka Maru was made by adding Norwegian whaling equipment to a ship body made in Japan (Tōyō Hogei K.K. 1910: 189–90). But Hōka Maru hardly caught any whales because the ship was slow and, apart from the Norwegian harpoon gunner, none of the crew had had any experience of Norwegian-style whaling. So, even at this early stage the company suggested having a ship built in Norway to use as a model for making ships in Japan (Matsu 1901a: 1901b, 1901c, 1901d, 1901e). Further, the whale catcher Hatsutaka Maru used by the Yamanobe group for coastal whaling off the Korean peninsula from 1901, was made by the Nagasaki Hogei K.K. from a Japanese body and whaling equipment imported from Norway. The Hatsutaka Maru also failed as a whaling vessel due to structural faults, so the Yamanobe group purchased a whaling ship from Norway (Tōyō Hogei K.K. 1910: 228–30).

11 But Yamada Tōsaku who was a founder of Nihon En'yō Fisheries and later became the controller of Tōyō Whaling, appears to have provided capital for net whaling. Incidentally, Oka Jūrō's family ran a sake brewing business (Fukumoto 1978: (1993) : 222; Tōyō Hogei K.K. 1910: 18, 190, 196–7).

12 Park takes the Chinese characters to be the Korean names Kim and Yu (Park 1995: 287), but it is possible that they are the Chinese names Jin and Liu.

13 After the flensing ship Mikhael was lent to Tōyō Whaling by the Ministry of Agriculture and Commerce as 'an experimental whale processing vessel,' from May 1906 it was anchored in Ayukawa harbour in Miyagi prefecture. But, for its time it was too gigantic to be of practical use. Although it was used as offices and as accommodation for company employees, it seems it was never actually employed in whaling *per se* (Kondō 2001: 230, 232; Tōyō Hogei K.K. 1910: 187, 236–7). According to the *Nihon Hogeigyō Suisan Kumiai Kikanshi* (the History of the Japanese Union of Fisheries and the Whaling Industry), it was later repossessed from Tōyō Whaling by the Japanese government in April 1913 and immediately sold to the Ishikari coal company (Ishikari Sekitan K.K.) (Park 1995: 283–5).

14 However, Russian Pacific Whaling's canned salted whale meat is said to have sold in Nagasaki for four yen fifty sen for a six hundred gram can

(Chōsenkai Tsūgyo Kumiai Rengōkai 1902b: 33). Further, the reason the position of salt store worker does not appear under 'nationals of this country' is presumably because this operation had been shifted to an on-shore facility.

15 Matsu Makisaburō, who had boarded the En'yō Hogei vessel, Hōka Maru, to observe the whaling activity, had argued that to prepare 'Japanese-style whale meat for consumption,' (Matsu 1901d: 17) it is necessary to bleed the meat in the same way as the *hazashi* do (Matsu 1901a: 19–20, 1901d: 17). He also insisted that *hazashi* were necessary because of their superior abilities, in comparison with the so-called seamen, in spotting whales and in steering boats to the spot where submerged whales were likely to surface (Matsu 1901d: 17). Actually, the Hōka Maru had two *hazashi* on board at the time Matsu was voyaging on the vessel (Matsu 1901a: 12).

16 In this company, thirty-one Japanese were employed as general process workers (Abe 1908b: 13; Tōyō Hogei K.K. 1910: 159). Moreover, in Emi's record it mentions that after a whale was taken they cut off the tail flukes to cut down on the effect of the waves (Emi 1907: 56–7)and this also seems to have been the practice with this company (Abe 1908a: 14).

17 In the case of Russian Pacific Whaling, oil was extracted by boiling the blubber in cauldrons installed in the flensing ship, but with Nihon En'yō Fisheries and the Holme Ringer Company the oil was extracted after transporting the blubber to the companies' respective plants in Japan.

18 There is visual evidence of this in *Hizenshū sanbutsu zukō – kodomo no moteasobi kujira ikken no maki* (Illustrated account of the products of Hizen province – the case of the child playing with the whale) (Kizaki 1970 [1773]: 822); *Ogawajima keigei gassen* (The battle with the whales at Ogawajima) (Hōshūtei 1840: (1995) 340–1); *Isanatori ekotoba* (Illustrated tales of whaling) 1832 [1829] ([1970]: 300)) and Shiba Kōkan's *Saiyū ryodan* (Traveller's tales of a journey to the west) (1794) and *Kōkan saiyū nikki* (Kōkan's diary of a journey to the west) (1815) (Torisu 1999: frontis piece illustrations 7 and 8). However, *Kumano Taijiura hogei no hanashi* (On whaling in Kumano's Taiji bay) (1937) (Kumano Taijiura hogeishi hensan iinkai ed. 1969: 444) and *Tosa Muroto Ukitsugumi hogei jitsuroku* (Actual whaling records of the Tosa Muroto Ukitsu group) (Yoshioka 1938: Figure 8, 16 (1973b): 412, 440) record that only after turning the whale so the head was pointing out to sea and the tail towards the beach was the animal dragged ashore.

19 This can also be seen in the sources listed under note 18 of this chapter.

20 In this source material, in the column for other prefectures, there are entries for 'fishing,' 'seaman' etc., but no details are given of the type of work performed or the employing company. And in the column for Iwate prefecture headed 'numbers of Koreans by occupation' there is a record of eleven employed in the 'fisheries salting industry.' It is assumed that this is probably the number of Koreans working as labourers for whaling companies (Naimushō Shakaikyoku Daiichibu 1924: 537–40). In addition, while, at the time, Korean workers' wages were in general a good twenty per cent lower than those of *naichijin* ('home islanders') (Naimushō Shakaikyoku Daiichibu 1924: (1975) 496–529), in this work the wages and conditions were the same for both groups. For information on the location of workplaces, see under 'investment data' in the Tōyō

Hogei column in the article, 'Tōyō Hogei K.K. nijūyoman shūeki zōka yosō' (Over twenty thousand yen predicted increase in profits for the Tōyō Whaling Company), in number 135 of the Ōsakaya Shōten Junpō (Osakaya Company Fortnightly Reports), 1926: 8–10.

21 It is thought that in the case of Tosa, most of these people must have been absorbed into bonito fishing and other coastal fishing industries, which had expanded greatly with the introduction of powered vessels (Izukawa 1942: 637–8 (1973c: 669–70). In Taiji, while the primary reason was the decline of net whaling prior to the introduction of Norwegian-style whaling, there was also considerable overseas emigration to North America and elsewhere (Hara 1993: 215–17; Ishida 1978: 99–104; Kumano Taijiura hogeishi hensan iinkai 1969: 351–3, 533). In this regard, it should be mentioned here that there were many engaged in the net whaling industry who opposed the introduction of Norwegian-style whaling, fearing that it would create mass unemployment and put their industry in jeopardy and in doing so they missed out on any opportunities the new technology may have provided (Hidemura 1952b: 105–6; Torisu 1999: 349–50; Tōyō Hogei K.K. 1910: 190–1). If we follow the argument of E. P. Thompson cited in note 8 of Chapter 2, perhaps this can be interpreted as a case of the 'moral economy' resisting the 'free market economy.' And one of those companies that missed out on the introduction of new whaling technology, the Ogawajima Hogei K.K. (a whaling group that changed to a company structure), operated in this area alongside Norwegian-style whaling (see Figure 1.1) for just the two years 1910 and 1911, after which it continued to exist in name only as the holder of the local whaling rights. From 1919 onward there was sporadic Norwegian-style whaling in the area and in 1941 there is an interesting report of collaboration between the Ogawajima Hogei K.K. and the whaling division of Hayashikane Shōten, in which a whale, spotted by a *yamami* (mountain lookout), was dragged onto the beach using a pulley and dissected by a few employees of Hayashikane Shōten's whaling division together with local residents (Kondō 2001: 316–19).

22 This technique had already been adopted by the Vancouver Island Whaling Company mentioned above.

23 Whale oil is divided into two main types depending on whether it is extracted from baleen whales or toothed whales. Baleen whale oil was used as the main ingredient in margarine and was also a source of glycerine used in the manufacture of explosives. Toothed whale oil, on the other hand, produced lubricating oil for industrial and military purposes (Baba 1942: 278–87).

24 The International Whaling Agreement (below in this note abbreviated to the Agreement), signed at an International Whaling Conference convened to conserve whale stocks and stabilise whale oil prices, formally came into effect in May 1937. Japan sent an official delegation to participate in the meeting the following year, 1938, but merely signed the protocol, indicating its preparedness to join the Agreement in 1939. At the 1939 meeting, which was not open to the public, Japan officially signed the revised agreement. But since World War II broke out during the process of ratification of the Agreement by those countries that had signed the revised document, the Agreement itself was declared null

and void (Ōmura, Matsuura Yoshio et al. 1942: 266–303; Baba 1942: 308–21).

25 Yamashita Shōto claims that it is misleading to keep pointing out that 'Japanese whaling was always consistent in its affirmation of the principle that whale meat production was the first priority, that oil production was of secondary importance and that when low priority was placed on whale meat production it was an exceptional circumstance that only applied to factory-ship whaling in the Southern Ocean' (Yamashita 2004b: 217). To support his assertion Yamashita points to the fact that after the outbreak of WWII, when it was no longer possible to export whale oil to Europe, whale meat production to secure food supplies also became a major of thrust of Antarctic factory-ship whaling. Conversely, he adds, during the Edo period (early 17th to late 19th centuries) it was not whale meat, but whale oil, as a stable commodity which could be kept for long periods, that became a mainstay of the economy (Yamashita 2004a: 235, 2004b: 217). Yamashita does not give his source for this latter claim, but we can find supporting evidence in Torisu's research (Torisu 1999: 91–152) in what is believed to be an account book of the Masutomi whaling group. If we limit our inspection to those sources cited by Torisu we would conclude that Yamashita's claim is correct. However, there is a need to investigate more closely whether or not the Japanese whaling industry has indeed always held that whale meat production is a higher priority than whale oil. In any case, the point I wish to make here is that both whale meat and whale oil were important products for the Japanese whale industry. Further, as Yamashita himself says, at the time of the introduction of Norwegian-style whaling, after defeat in the Pacific War and since 1960, whale meat production has been the main focus of factory-ship whaling in the Antarctic (Yamashita 2004b: 175, 217, 240–4, 258–65). Whether meat or oil production was given priority depended on the various conditions prevailing at the time.

26 From the context we can assume that this figure of 300 refers to those in the Nihon Suisan–Nihon Kaiyō Gyogyō Tōsei K.K. (the Nihon Fisheries–Nihon Kaiyō Fisheries Control Company). And given the circumstances of the time we would guess that these were general seamen and process workers.

Chapter 2

1 One could argue against this along the lines of E. P. Thompson's criticism of earlier research in his *The Making of the English Working Class*. In his analysis of Luddism (the breaking of machinery by workers in the textile industry) in the early nineteenth century, Thompson argues that if we assume that Luddism was fanned by agents of the authorities, we cannot explain its subversive nature or fit it into the context of the local community where informers were ostracised by the whole village (Thompson 1980: 684–716). We cannot show how Luddism was subjective action taken by the common people, nor describe the popular world it had spawned.

2 Political opportunity structures, to take the definition of S. Tarrow as it is quoted in Nakazawa et al., form 'the political context of movements, including those factors that suppress activity and those that provide

incentives for collective action, by influencing the expectations of actors regarding the success or failure of the movement' Nakazawa et al., 1998: 144. In other words, they may be considered as the circumstances and networks involved when social movements take political action. Their constitutive elements can be divided into, fixed systems (councils, participatory citizens institutions etc) and fluid structures (policies, political discourse, and alliances between political parties and citizens etc) and the notion of political opportunity is said to explain when a movement will occur (Takubo 1997: 135–9).

3 Resource mobilisation refers to the financial and material resources and so-called human resources, plus such things as networks of human relationships that are mobilised in social movements.

4 This definition of framing is similar to A. Melucci's concept of collective identity, i.e. 'an interactive process through which several individuals or groups define the meaning of their action and the field of opportunities and constraints for such an action'(Melucci 1996: 67). There is a reason for this. Melucci makes the following criticism of resources mobilisation theory (RMT) as it has developed under the formularisation, mentioned above, of some of the theory's advocates. 'Its key concepts, such as "discretional resources" and a "structure of opportunities," do not refer to "objective" realities but imply the capacity of actors to perceive, evaluate and determine the possibilities and limits afforded by their environment. RMT thus postulates a process of actors' construction of their identity, without, however, examining this process' (Melucci 1989: 34). This criticism is representative of those that decry RMT's lack of concern for socio-psychological factors. The concept of framing, we can assume, was introduced by those who had been responsible for the formularisation referred to above, in order to respond to this line of criticism.

5 For example, Takubo Yūko analyses the three elements mentioned above in her research on the movement dubbed the 'Referendum Advocacy Association' of Maki town in Niigata prefecture, which in 1996 successfully lobbied for residents to be given a vote over the introduction of a nuclear power station into their town. Takubo contrasted the so-called anti-nuclear movements with the 'Referendum Advocacy Association.' She has shown that the association contributed to the formation of a frame of reference according to which there can be movements that demand the 'execution of the public will' in the literal sense of the phrase – as distinguished from movements that develop specific political claims. She has also demonstrated that the association promoted the principle that 'it is democracy to let the people decide' and succeeded in gaining media attention by actively demanding 'disclosure of information' (Takubo 1997: 142–3). But very little is said about how the members of the 'Referendum Advocacy Association' came to reach shared understandings or subsequently how those understandings developed.

6 Melucci cited in Note 4 above seems if anything to be more interested in 'internal framing.' This is clear from the fact that, under the influence of Touraine's 'sociological intervention' method, he has become actively involved with groups engaged in social movements and has carried out empirical research to discover within these the processes that give rise to collective action (Melucci 1989: 197–204, 235–59). Further, my own

attention to 'internal framing' stems from an awareness of Melucci's arguments regarding contemporary social movements. Further, to give a brief explanation of Touraine's 'sociological intervention' method, it runs in the following stages: (1) 'An intervention group' composed of militants enters confrontations with a plurality of interlocutors (supporters of the movement, adversaries and specialists), (2) A group of researchers, divided into those taking the roles of agitators and those who are secretaries, instigate the members who make up the intervention group to analyse the conditions and meanings of their own actions, (3) 'conversion' – the research group proposes a hypothesis and communicates to the 'intervention group' an interpretation of its actions that reflects the highest possible meaning and strives to understand the action from this perspective, (4) 'permanent sociology' – the research group prepares a report of its first round of conclusions and this is debated between the intervention group and other groups of activists. In this process the research group focuses on the capacity of, and methods used by, the actors to resolve the analysis gained through the intervention into a programme of action. For a detailed explanation see (Touraine et al.1980, Japanese translation, Itō Ruri [1984], *Hangenshiryoku undō no shakaigaku*, Shinsensha).

7 For example, Yoshida Ryūji in his analysis of 'crowd-specific interaction' (*gunshū tokuyū no sōgo sayō*) in the First Kamagasaki Riot of 1961, points out the 'continuity of crowd behaviour and institutional behaviour' (Yoshida 1994). In this, Yoshida claims that there is a close connection between the behaviour of those participating in the riot and their everyday 'world of meaning' (*imi sekai*), i.e. 'all the accumulated meanings each member of a certain social group applies to his or her environment (i.e. the social entity of the self and all social and physical relationships). Or, those among the individual meanings that each member of that group applies to his or her environment that can be regarded as being shared by all in the group. These are regarded by the outside observer as the fundamental attributes of a certain social group' (Yoshida 1994: 84). And he clearly shows how the behaviour takes into consideration the ambivalent feelings of 'fear' and 'pity' that 'society at large' has towards Kamagasaki, and that 'society at large' is not merely a passive 'audience,' but performs the role of a medium, and sometimes even provides a solution, for the behaviour of the rioters. Further, Matsuda Motoji focuses on everyday practice, which he sees as both the source of a certain creativity and a combination of 'tradition' and 'resistence' (Matsuda 1996; 1997). This notion is akin to de Certeau's idea of a 'tactic,' i.e., to quote from Steven Rendall's English translation, 'a calculus which cannot count on a "proper"(a spatial or institutional) localization, nor thus on a borderline distinguishing the other as a visible totality.' ... 'because it does not have a place, a tactic depends on time – it is always on the watch for opportunities that must be seized "on the wing"' (de Certeau 1980: 26, Rendall trans. 1984: xix). According to Matsuda the idea of power today is shifting from macro, hard and homo[geneous] to micro, soft and hetero[geneous] forms, which make eternal control possible. Further, he maintains that accompanying this trend we can now also see 'resistance' in terms of micro, soft and hetero. In other words, rather than organised, violent 'resistance,' the

focus of attention is now on 'resistance' centred in the practical minutiae of everyday life (micro), steeped in the passive and submissive (soft) and diversified beyond the monolithic aggregate of the oppressed (hetero).

Matsuda asserts that the kind of resistance he himself calls 'soft resistance' and defines as 'the process of making use of the laws and norms forced upon us, to create a variety of alternatives not envisaged in the original plan' (Matsuda 1997: 123) can be identified even within riots. In short, he claims that, even in a riot certain values and norms are created that are a 'continuation of the tradition that underpins the everyday world' (Matsuda 1997: 129).

In addition, Melucci points out that the basis of 'collective identity' is formed by a 'network of affiliation' to organisations, whether they be community associations based on the village or among people considered to be of the same ethnic group, or special interest associations formed around hobbies and the like, that existed before the collective action arose (Melucci 1996: 289–92).

8 Thompson to whom I referred in Note 1, points out the following two points as the cause of Luddism. Firstly, the fact that at around this time the abrogation of paternalist legislation that made a fixed period of service as an apprentice compulsory (this had the result of inhibiting the intake of unskilled workers), infringed the rights of the journeymen, which had been firmly rooted in the Law and custom (Thompson 1980: 594). Secondly, the value system and lifestyle of factories that used machines encroached on the values and lifestyles of the journeymen and their communities. To put this second point another way, they were resisting the introduction of the 'laissez-faire economy' that resulted in the lowering of wages and a deterioration in quality, since what had been previously made by journeymen could now be produced with simple labour in factories. Their opposition was based on what Thompson calls the 'moral economy,' which, by placing appropriate value on the products of their labour, paying fair wages and through the practice of mutual help was able to sustain the journeymen throughout their lives (Thompson 1980: Japanese translation 2003: 619–56). Matsuda, cited in Note 7 above, mentions 'the moral economy,' referring to Thompson's research other than in *The Making of the English Working Class*, in arguing his point, alluded to earlier, about the revitalisation of 'tradition' in riots (Matsuda 1997). But in the context of the formation of the English working class, in the light of the various intertwining factors hinted at by Thompson, such as the works of Thomas Peine and the influence of the Methodists, I feel there is still room for argument over claims that the moral economy was the enduring essence of the journeymen and the working people and that it provided the foundation for the formation of the working class.

9 The Preliminary Hearing refers to the investigative procedure, stipulated under the old Code of Criminal Procedure, in which, after the charges have been filed, a magistrate decides whether a case should go to trial. At the same time it provided an opportunity for the gathering and preservation of evidence that might be difficult to investigate in the public trial. Proceedings were held in camera and no lawyers were present. Moreover, the Report of the Preliminary Hearing which

recorded the magistrate's findings had unconditional admissibility as evidence in the court of the public trial (Suekawa 1978: 1005).

10 Both rolls of microfilm had a few sections in which the text was unclear due to mistakes in the photographing process. One of the librarians at the Hachinohe City Library explained that this was because the microfilms had been made by library staff rather than specialist professionals.

11 We can see from the transcript of the sentences (See Note 14), that one of the men of influence who had encouraged the introduction of the whaling station and whose house had been attacked, gave evidence at the preliminary hearing similar in content to that quoted here from the determination of the preliminary hearing (Satō 1987: 98). But, possibly because the sources had been lost, this man's testimony cannot be confirmed from the transcript of the interrogation of witnesses included in Micro. A.

12 In regard to participation in the riot, it was claimed (in the testimony of defendants K, N and O on Micro. B) that they had agreed on the rule that there should be at least one person from each household and that anyone who failed to take part should be prepared to accept sanctions ('Same bōdō yoshin shūketsu,' 1911b). When considering mobilisation, then, it appears we can cite the fact that everyday social ties within the livelihood organisations played a specific role in this incident.

13 The widespread protest movements against the introduction of Norwegian-style whaling mentioned at the beginning of this chapter also base their opposition on the fact that whales drive fish closer to shore making them easier to catch and the blood and oil from the flensing of whales is harmful to the marine environment. Tōyō Whaling dismissed these claims as 'superstition' ('Dotō no hibiki,' Dai Nihon suisankaihō, 339, pp. 32–3; Matsuzaki 1910; Tōyō Hogei K.K. 1910: 242–4).

14 For the sentences I have consulted Satō's book which has the full text in what appears to be a modern language translation, rather than referring to the text of the sentences in Micro. B.

15 Articles dealing with the trial appeared in the *Hachinohe* and the *Ōnan Shinpō*, but, apart from a few days of the trial, the contents were identical. As it seems likely that both the *Hachinohe* and the *Ōnan Shinpō* were quoting from the *Tōō Nippō*, which was first into print with news of the trial, I have referred to the *Tōō Nippō*, except for those days of the trial when the content of the articles differed.

16 After the questioning, the Court read the transcripts of their testimony to the defendants and other witnesses and required them to acknowledge the accuracy of the reports by signing and affixing their seals to the documents. The record shows that the clerk of the court signed on behalf of five of the twenty-nine defendants listed in Table 2.1, because they were unable to write. Moreover, the fact that some transcripts lacked seal and signature suggests that perhaps the actual number of illiterates was somewhat greater.

17 Ishida Yoshikazu quotes from this document in his book, but he does not clearly identify the source so it could not be brought out in the process of the investigation.

18 In the end, whaling operations started up in Chōshi, Ushitsu and Same, the places I mentioned at the beginning of this chapter. But this soon

caused a reduction in the numbers of those comparatively larger species of whale that were considered useful as a 'resource' (Okada 1916) and by 1926 at the latest, operations had ceased to be viable at Chōshi or Ushitsu. In addition, there was a gradual decline in catches in those areas in Honshū and Kyūshū where whaling had formerly been practised by whaling groups (*kujiragumi*). Consequently, the Japanese whaling industry sought to establish stations not only in other areas on the Korean Peninsula, but in eastern Hokkaidō, the Kurile Islands (Chishima) and Southern Sakhalin (Karafuto) (Nōrinshō Suisankyoku 1939: 5–9); (Ōta 1927); (Tōyō Hogei K.K. 1910: Contents 9, 19–20).

19 The figures were 29 whales in 1947, comprising 13 fin whales, 6 sei whales and 10 sperm whales; 14 (6 fin, 4 sei and 4 sperm) in 1948 and 8 (3 fin, 4 sei and 1 sperm whale) in 1949.

20 In an article on whaling in the Antarctic Ocean with the Nisshin Maru fleet in the 1956–57 season, the *Yomiuri Shimbun* reported that 132 of the 640 workers engaged in processing aboard the factory ship Nisshin Maru and its one freezer ship came from the Hachinohe area of Aomori prefecture ('"Nanpyōyō imin" no kiroku,' *Suisan Kikan* (Fisheries Quarterly), 4, pp. 96–9). We can assume that this indicates that after the closing of the Same whaling station, those who had worked there, or people from the area influenced by the fact that there had been a whaling station at Same, came to become engaged in factory-ship whaling in the Antarctic Ocean.

Chapter 3

1 For references to *Kanpō* (The Official Gazette) and *Chōsen Sōtokufu Kanpō* (Official gazette of the Government-General of Chōsen) the issue number and date of publication are given.

2 See the author's discussion of the place of enactment and designation of natural monuments in Watase's thought overall (Watanabe 2000).

3 In 1915 Watase travelled to Canada and the United States of America to inspect fox farming for the fur industry (Yatsu 1931: 46; Watase 1916a; 1916b; 1916c; 1916d; 1916e; 1916f; 1916g).

4 The Mammalogical Society of Japan has established a ranking of five stages of endangerment from most to least endangered, 'Extinct,' 'Endangered,' 'Vulnerable,' 'Rare' and 'Common.' (In addition it has the categories 'Insufficiently known' and 'Common as a whole, but locally includes threatened population(s)'). For a detailed description see (Nihon Honyūrui Gakkai 1997: 9–11).

5 However, although the breeding grounds of the black-tailed gull on Kabushima and the streaked shearwater of Ōshima were designated natural monuments in 1922 and 1928 respectively, the breeding grounds of the streaked shearwater of Birōjima were not so designated.

6 Even Kaburagi's report acknowledges that the finless porpoise lacked commercial value, its only practical use being in providing oil used for the eradication of insect pests.

7 Guano is the solidified accumulation of seabird droppings used as fertilizer.

8 Japanese translation, Maehata Masayoshi et al. [1996], *Kujira to iruka no zukan*, Nihon Bōgusha, pp. 50–3.

9 In regard to their relationship with laws in the so-called 'home-islands,' the status of laws in the colonies differed according to the historical circumstances of colonisation in each of the colonies. In the case of Chōsen, fundamentally, there was no right of representation in the Imperial Diet (the right was recognised in January 1945 when consideration was given to the introduction of conscription in Chōsen (enforced in April that year), but in fact no election of councillors to the House of Representatives was ever held) and there was no local assembly. Accordingly, the majority of laws took the form of government ordinances, presented through the prime minister for imperial approval and then enacted as a directive from the Governor-General of Chōsen (Kajimura Hideki and Kang Duk Sang 1970; Nakamura 1958).

10 On whaling regulation refer to works of Shigeta, Baba and Ōmura et al.(Shigeta 1962a; 1962b; 1962c; 1962d; 1962e; Baba 1942: 301–26; Ōmura, Matsuura Yoshio et al. 1942).

11 For a discussion of whaling regulation in Chōsen under colonial rule see Park Koo-Byong(Park 1995: 278–81).

12 For reference to the *Chōsen Sōtokufu Kanpō* (Official gazette of the Government-General of Chōsen) I have used the facsimile edition which is missing some issues for 1944 and 1945.

13 At this time small-type (coastal) whaling in the so-called 'home islands' (*naichi*) did not fall within the Ministry of Agriculture and Forestry's range of permitted fishing (see Chapter 1). But, as an emergency measure, from 15 January 1944 for a limited period of one year, the whale catchers that had been engaged in small-type coastal whaling were given special permission to operate, within the boat limits for coastal whaling, in order to increase food production and train seamen, both of which were in poor supply. The ships were brought into the whale catcher quota system and five vessels were allocated to each of the three whaling companies, Nihon Kaiyō Gyogyō Tōsei K.K. (Japan Ocean Fisheries Control Company), Nishi Taiyō Gyogyō Tōsei K.K. (Western Ocean Fisheries Control Company) and Ayukawa Hogei K.K. (Ayukawa Whaling Company)(Maeda and Teraoka 1952: 33–4; Shigeta 1962e: 10).

14 This regulation can be regarded as a direct continuation of a clause in the Whaling Industry Management Law issued by the imperial Korean government in 1907 (Park 1995: 266–7; Tōyō Hogei K.K. 1910: appendix 5–6). Incidentally, as a result of the (second) Japan – Korea agreement of 1905, Korea became a Japanese 'protectorate' and a Resident General's Office was established.

15 I was able to confirm the subsequent amendments and content of the Management Regulations for the Protection of Chōsen Fisheries by checking government ordinances and notices in issues of the Chōsen Sōtokufu Kanpō (Official gazette of the Government-General of Chōsen) up to August 1945.

16 I was unable to find the 'Schedule of Categories for the Conservation of Treasures, Ancient Sites, Scenic Places and Natural Monuments in Chōsen' in the *Chōsen Sōtokufu Kanpō*.

17 Incidentally, the actual category under which the grey whale was designated a natural monument is unclear.

18 For the following description of damage to agriculture and forestry and the number of deer taken, I have used the material published on the website of the Natural Environment Division, Office of Environment Affairs, Department of Health and Environment, Hokkaidō Government, under the title 'Ezoshika no hogo to kanri' (The conservation and management of Hokkaidō *shika* deer), *http://www.pref.Hokkaidō.jp/kseikatu/ks-kskky/ sika/si katop.htm* (accessed on 22 September 2004). Refer to this website for recent developments in the conservation and management of *shika* deer in Hokkaidō.

19 As there is no domestic market in Japan for venison, those who advocate the 'effective use' of the Hokkaidō *shika* deer have been making trial calculations with a view to exporting the meat to Germany where there is a stable market. To transport the meat from eastern Hokkaidō to Germany costs ¥200 per kilogram (¥100 for sea freight in 5 ton lots). The cost of butchering within Japan to EU standards, including the cost of processing to adhere to the stipulated time limit for unskinned deer carcasses of within seventeen days from the time of death, comes to ¥2,000 per kilogram. With the transport cost added in, the figure becomes ¥2,200 per kilogram. This price is excessively high compared to the price of farmed deer meat from New Zealand, which, after export to Germany, fetches a purchase price (it is not clearly indicated whether this is the purchase price, but this seems to be the case, judging from the context) of ¥600 per kilogram (Ōtaishi and Honma 1998: 190–5).

Chapter 4

1 I am indebted to Mr Kondō Isao for pointing out this fact.

2 I have corrected the average figures here for the years 1849 to 1865, as the results recorded in the source materials seem to be incorrect. Averages are rounded to two decimal places.

3 According to Maeda Keijirō and Teraoka Yoshirō (Maeda 1952: 171, 174–5). But Tōyō Whaling's edited volume (Tōyō Hogei K.K. 1910) carries an advertisement for the Isana Shōkai between pages 36 and 37 that says, 'major sale under exclusive contract to sell Tōyō Whaling's whale meat east of Osaka.' Further, Yamada Tōsaku (Maeda 1952 has Yamaguchi Tōsaku on page 174, but I take this to be a misprint) was, in 1910, an inspector for Tōyō Whaling (Tōyō Hogei K.K. 1910: 18, and the caption to the photograph following page 80).

4 According to the report in the *Tōkyō Nichinichi Shinbun* 'Geiniku o uru' (Whale meat to go on sale) (1919), at that time red whale meat was priced at 16 sen per one hundred monme (375 grams), and white meat (of the ventral pleats) was 30 sen, in contrast to beef sirloin at one yen sixty sen, chicken breast at one yen eighty sen and horsemeat at sixty sen. It is not clear, however, whether these whale meat prices reflect the usual market price or whether they were discounted to below cost for the sale.

5 Maeda and Teraoka give 1908, the year of the trial production at Tōyō Whaling's Ayukawa station in Miyagi prefecture, as the beginning of canned whale meat (Maeda 1952: 173), but this seems to be a mistake.

Andō Shunkichi's observation in 1913 that 'it started over ten years ago' is probably closer to the mark (Andō 1913: 14).

6 The basis for this assertion seems to be Maeda and Teraoka's claim that 'the manufacture of canned whale meat began early in the Tōhoku district' (Maeda 1952: 175), but in the light of note 5 above, this is questionable.

7 To confirm and correct discrepancies between the table and the map, I referred to Jinbunsha's *Nihon bunken chizu chimei sōran* (A compendium of Japanese maps and place names by prefecture) (Henshūbu, Jinbunsha ed. 1997), and Nihon Kajo Shuppan's *Zenkoku shichōsonmei hensen sōran* (A compendium of city, town and village name changes) (Nihon Kajo Shuppan K.K. ed. 1979).

8 Izukawa himself states that his selection of two metropolitan areas (fu) and thirteen prefectures (ken) in the Kinki and Chūbu districts is entirely arbitrary.

9 In this connection, the number of chickens in Aichi prefecture in 1941 was 6,222,216, or 16% (rounded to the nearest whole number) of the national total (39, 591,617 excluding the colonies). This was the highest number of any prefecture and considerably above the second largest producer, Chiba prefecture, with 1,722,701 birds (Nōrin Daijin Kanbō Tōkeika 1942: 367).

10 As at 1939 both Nihon Suisan K. K. (the successor of Tōyō Whaling) and Hayashikane Shōten based themselves in Wakayama prefecture (Nōrinshō Suisankyoku 1939: 5–6).

11 Nothing in particular happened fifty years ago, but it is possible that the response 'about ten years ago' might include the momentum of Japan's waging of war from the Manchurian incident on.

12 There were 19 communities on the map (2 in Kyoto; 3 in each of Wakayama, Mie, Shizuoka and Toyama; and 5 in Ishikawa) and 14 in the table (2 in each of Kyoto, Mie, Toyama and Niigata; and 3 each in Shizuoka and Ishikawa) that answered that they made use of whale meat obtained locally. Of the communities in the table the following five indicated their method of catching whales: Shinminato in Toyama prefecture – driven into yellowtail set-nets (*buriami*); Kuki and Shimazu in Mie Prefecture – captured in triangular set-nets (*ōshikiami*); Shimazu, Mie prefecture, Chihama, Shizuoka prefecture and Sotokaifu, Niigata prefecture – utilise (utilised) stranded and drifting whales. Interestingly, both communities in Niigata are on Sado Island. Further, there were responses from six locations on the map, from the tip of the Nōto peninsula to the Arisoumi area of Toyama prefecture, which are known for whaling using the set-net technique (*daiami*) (Fukumoto 1978: (1993): 41–2, 52–4; Kitamura 1838: (1995): 132–4, 170–81).

These six places that replied they made use of whale meat acquired locally, included Ushitsu (recorded in the table written with two characters and not three, but I have assumed this is a misprint) and Ogi, both in Ishikawa prefecture and both locations where whaling companies had established whaling stations some time after (the exact date is unclear) the advent of Norwegian-style whaling (Tōyō Hogei K.K. 1910: 20 and Chapter 2, above).

13 Actually, of the communities that replied they had begun eating red whale meat within the past ten years, there were three in Kyoto (which indicated

respectively that they had been eating blubber, 'for fifty years,' 'since before the Meiji period' and 'from a very long time ago.' Further, there were two communities in Osaka that responded that they had only begun eating *akaniku* very recently and that in the past no raw whale meat came into the village. The former claimed they had been eating whale meat 'since the old days' and the latter 'from about fifty years ago.' Moreover, all communities in the Osaka–Kyoto area that replied that whale meat had been introduced between thirty-one and fifty years ago, were now using *akaniku*.

14 In this connection, of the five prefectures (here the term includes the metropolitan divisions, *fu*) that indicated a relatively low level of whale meat consumption, twelve communities indicated they ate only red meat and two only canned whale meat. Of those communities that claimed whale meat had been introduced before Norwegian-style whaling (more than fifty years ago), two responded that they ate only red meat and none claimed to eat only canned meat. In those communities where the date of introduction of whale meat was unclear, three claimed to eat only red meat, while none ate only canned meat. For the land-locked prefectures, thirteen communities responded 'red meat only,' while nineteen claimed to eat only canned whale meat. In these prefectures, only one of the communities where whale meat consumption predated the introduction of Norwegian-style whaling, claimed to eat red meat only and none replied 'canned whale meat only.' Where the date of introduction of whale meat was unclear, two communities ate only red meat and twelve only canned whale meat.

15 In Izukawa's table, too, there are three examples (one in Kyoto and two in Niigata) where it is claimed that whale meat was (or is) eaten on New Year's eve (*ōmisoka* or *toshikoshi*).

16 From Nango in Yamanashi prefecture we have a report to the effect that 'it is mainly dolphin and the like, stewed and flavoured with miso bean paste...', indicating that the eating of dolphin meat was not restricted to the Izu peninsula. It seems it was distributed from that area to surrounding districts. Further, Nakamura Yōichirō also cites the *Shizuoka-ken Suisanshi* (A compendium of Shizuoka prefecture fisheries) of 1894 and the *Nihon Suisan Hosaishi* (A compendium of Japanese fishing methods and fishing gear) (year of publication not indicated) reports that dolphin meat circulated through Sagami (Kanagawa prefecture), Kōshū (Yamanashi prefecture) and Shinshū (Nagano prefecture), and a dried dolphin meat (*tare*) was sold in Owari-Mikawa (Aichi prefecture) and Mino (Gifu prefecture) (Nakamura 1988: 126–8). But dolphin fishing was never carried out on the same scale as whaling and even on the Izu peninsula dolphin drives, it seems, reached their peak from WWII until around 1965–75 (Nakamura 1988: 104–17; Shizuokaken Kyōiku Iinkai Bunkaka 1986: 44–5, 105–27, 168–9; 1987: 69–97,182–3), so before this time we can consider that dolphin meat was not eaten on a daily basis outside the Izu peninsula and its immediate vicinity.

17 In Tosa (Kōchi prefecture), where net whaling had been practised, those engaged in catching activity within the whaling groups became involved in bonito fishing in the summer, when whales were not hunted. Here, both the sardines, the bait for bonito fishing, that came in with the whales, and the bonito that came in pursuit of sardines, were called *kujirako* (whale

children) (Yoshioka 1938: 52–3; 1973b: 476–7). That is to say, the people
in the Tosa *kujiragumi* (whaling groups) benefited from the blessings of
living whales in bonito fishing on the one hand, but hunted and used whales
on the other.

Chapter 5

1 The 1938 figure does not include the colonies. Further, Okinawa was not
included in the national statistics for Japan from 1944 to 1972 and the
Amami Islands were excluded between 1945 and 1953 (Nōsei Chōsa Iinkai
1977: iii). Consequently, the 1946 statistics exclude both Okinawa and the
Amami Islands and the figure for 1956 excludes Okinawa. Also, it seems
likely that the 1946 statistics do not include the Ogasawara (Bonin) Islands
and it is possible that the 1956 statistics do not include the Amami Islands.
Incidentally, the peak of pig production excluding Okinawa was also in
1938 when 998,985 head were bred. Similarly, it was not until 1956 that
this peak for pork production excluding Okinawa was finally exceeded.

Taking these facts into account, in the discussion of livestock statistics
in this chapter, for my assessment of what constitutes 'the home islands'
(*naichi*), I have interpreted terms in the source documents such as *waga
kuni* (our country), *kokunai* (within the country) etc. as being equivalent
to the scope of the present Japanese prefectures (*todōfuken*). However, it
is possible that *waga kuni*, *kokunai* etc. do not include Okinawa.

2 The number of whale catchers permitted in the so-called home islands
was thirty from 1909, reduced to twenty-five in 1934 (See Chapters 1 and
3). However, it seems that ships could be substituted for licenced vessels
and, after the beginning of Japanese whaling in the Antarctic, training
ships were permitted for the instruction of harpoon gunners and crew.
Consequently, the number of whale catchers actually operating was larger
than these figures would suggest (Maeda and Teraoka Yoshirō 1952: 98–9).
This is why larger numbers of operating vessels are cited in some accounts,
e.g. (Freeman 1988: 16; Nagasaki 1984: 78).

3 With the introduction of a licence system, operations which had previously
been carried out from vessels called *minkusen* or *tentosen* came to be
known as small-type whaling (Maeda and Teraoka Yoshirō 1952: 117).
Further, we can assume that when small-type (coastal) whaling came
under a permit system, the previous general coastal whaling industry
came to be called large-type (coastal) whaling (Kanpō (Official gazette)
1947, 6269).

4 It has been pointed out that this had been set up to get around the
strengthened restrictions on factory-ship whaling in the Antarctic Ocean
(Hara 1993: 115–19). From 1963 Taiyō Fisheries, Kyokuyō Whaling and
others had used the stations of a Dutch firm and Nippon Fisheries that
of a British company on the British territory of South Georgia Island.
They set out from these bases at the southern tip of the South American
continent to carry out whaling in the Antarctic Ocean ('Dai jūniji nangei
no shutsuryō keikaku' (12th (sic.) Antarctic Ocean whaling plan) (1964),
Suisankai (Fisheries world), 959: 54–6; 'Kankeikoku chūshi no naka o
Nihon sendan wa shutsuryō shita' (The Japanese fleet puts to sea under the
gaze of member countries) (1963), Suisankai (Fisheries world), 947: 50–1).

This whaling from land-based stations was ultimately unsuccessful, with Taiyō Fisheries and Kyokuyō Whaling withdrawing after the 1964 season and Nippon Fisheries pulling out in 1966, all having paid cancellation fees for terminating their contracts (Hara 1993: 125–6).

5 Baba says coastal whaling was prohibited from 1894, but the correct date is 1904 (Shigeta 1962a: 8, 15). Incidentally, one of the reasons why herring fishermen were opposed to whaling, was that they saw whales as 'heavenly envoys' (ten no shisha) (Sekizawa 1888: 22) that guide herring and other fish to shore. Sekizawa Akekiyo, writing in 1888, introduces this in the context of a Norwegian scientist's rebuttal of anti-whaling rhetoric in a lecture probably given the year before (1887), in which he refuted the belief in Hokkaidō that whales are manifestations of the god Ebisu (Sekizawa 1888: 21, 29). In addition, this period coincides with the decline of coastal whaling in Norway due to overfishing. This, along with the ban on coastal whaling described above, sent Norwegian whaling companies and harpoon gunners off in quest of new whaling grounds abroad. Yamashita Shōto points out that this was the principle reason why, at the time of the introduction of Norwegian-style whaling, it became possible for Nihon En'yō Fisheries to charter Norwegian whale catcher boats (See Chapter 1)(Yamashita 2004a: 164–6, 173–4).

6 Watase claims that, as at 1965, the sperm whale quota was set far below the maximum possible catch and for that reason it led to the suspicion abroad that Japanese sperm whale numbers had decreased markedly and resulted in the domestic industry becoming economically unviable. He goes on to say, 'as coastal whaling is run by private enterprise, in order to bring it a little closer to breaking even economically, a certain amount of what is popularly called "fuzzy accounting" is employed. (Watase 1965: 44).

7 This source is introduced by Hara (Hara 1993: 106–10).

Chapter 6

1 But we need to bear in mind that the belief in the god Ebisu (Ebisugami) takes a number of complicated forms. The title Ebisugami can be applied not only to living creatures such as whales, dolphins and sharks, but also to gods recorded in the ancient chronicles (the Kojiki and the Nihonshoki) dedicated in shrines called Ebisu shrines (Ebisu no yashiro), or to the corpse of a drowned person (Namihira 1978) and so on. Of the songs associated with the belief in Ebisu, one sung at banquets and as a work song of the Masutomi whaling group mentions Ise shirine and the divine favour of Ebisu (Morita 1994: 160–6). In Kawana on the Izu Peninsula there is an Ebisu service included in worship at the Ebisu shrine to a local god of fishing, on the anniversary of the day when it is said that Kawana first succeeded in a dolphin drive (Shizuokaken Kyōiku Iinkai Bunkaka 1987: 96). We should probably interpret these examples where a whale has been killed, or a dead whale has been used, for its meat and oil as a belief in the god Ebisu, not as the whale itself, but as the god that brought about the capture of the whale.

2 The author has worked mainly from the Japanese translation, Takahashi Jun'ichi et al. trans. [1989], *Kujira no bunka jinruigaku*, Kaimeisha.

3 Japanese translation, Takahashi Jun'ichi et al. trans. [1989]: 44.

4 Japanese translation, Takahashi Jun'ichi et al. trans. [1989]: 165–6.
5 Japanese translation, Takahashi Jun'ichi et al. trans. [1989]: 185–6.
6 Japanese translation, Takahashi Jun'ichi et al. trans. [1989]: 6–7.
7 In addition to these, Takahashi bases his continuity argument on the fact
   that 'traditional' religious beliefs in whaling remain and that memorial
   services for whales and associated rituals, either at the level of the
   whaling company or involving the whole of the local community, are still
   carried out today (Takahashi 1992: 116). But I have chosen to disregard
   this evidence here, because, although he mentions this fact, he gives no
   concrete examples.
8 For example, the governor of Tokyo, Ishihara Shintarō, a typical right-wing
   politician in present Japan, when questioned by Sataka Makoto over the
   strong advocacy of the 'mixed race theory' of Japanese origins, made in
   the past to justify Japan's policy of expansion in Asia under the slogans of
   'all the world under one roof' (*hakkō ichiu*) and 'harmony among the five
   races' (*gozoku kyōwa*), Ishihara answered unequivocally that the Japanese
   are a mixture of many ethnic groups. It seems his reply was based on his
   conviction that 'superior individuals' can emerge more readily from those
   of 'mixed blood' and that he was suggesting thereby that 'the Japanese are
   superior' (Ishihara and Sataka 2000). Be that as it may, what is important
   here is the fact that the hybridity view that the Japanese are of mixed
   ethnic origins, rather than being a mono-ethnic group, is to be found in
   the minds of some nationalists, both now and in the past.
9 And the Japanese translation, Takahashi Jun'ichi et al. [1989]: 12–13,
   39–40, 43–4.
10 Friedheim's position is to accept all coastal whaling, indigenous or not,
   and to have the IWC develop appropriate management plans, so that non-
   member countries are encouraged to join the IWC (Friedheim 2001: 315–16,
   324–7).

Appendix

See Table A.1

Table A.1: Glossary of Japanese Whaling Company Names 1897–1972

Japanese Name	Romanisation	Translation	Short Reference	Dates
有川捕鯨会社	Arikawa Hogeigaisha	Arikawa Whaling Company	Arikawa Whaling	1897–1900?
遠洋捕鯨株式会社	En'yō Hogei Kabushiki Kaisha (K. K.)	Pelagic Whaling Company	En'yō Whaling	1897–1900?
長崎捕鯨株式会社	Nagasaki Hogei K. K.	Nagasaki Whaling Company	Nagasaki Whaling	1897–?
ホーム・リンガー商会	Hōmu-Ringā Shōkai	Holme-Ringer Company	Holme-Ringer Co.	1898–1901
日本遠洋漁業株式会社	Nihon En'yō Gyogyō K. K.	Japan Pelagic Fisheries Company	Nihon En'yō Fisheries	1899–1904
山野辺組	Yamanobegumi	Yamanobe Group	Yamanobe Group	1901–1903
長崎捕鯨組合	Nagasaki Hogei Kumiai	Nagasaki Whaling Partnership	Nagasaki Whaling Partnership	1903–1904
大韓水産会社	Daikan Suisangaisha	Great Korea Fisheries Company	Daikan Fisheries	1904–?
長崎捕鯨合資会社	Nagasaki Hogei Gōshigaisha	Nagasaki Whaling Limited Partnership	Nagasaki Whaling Limited Partnership	1904–1909
日韓捕鯨株式会社	Nikkan Hogei K. K.	Japan-Korea Whaling Company	Nikkan Whaling	1904
東洋漁業株式会社	Tōyō Gyogyō K. K.	Oriental Fisheries Company	Tōyō Fisheries	1904–1909
大阪春日組	Ōsaka Kasugagumi	Osaka Kasuga Group	Osaka Kasuga Group	1906–1907
日諾捕鯨会社	Nichidaku Hogeigaisha	Japan-Norway Whaling Company	Nichidaku Whaling	1906–1910?
日韓捕鯨合資会社	Nikkan Hogei Gōshigaisha	Japan-Korea Whaling Limited Partnership	Nikkan Whaling Limited Partnership	1906–1919
太平洋漁業株式会社	Taiheiyō Gyogyō K. K.	Pacific Ocean Fisheries Company	Taiheiyō Fisheries	1907–1908
東海漁業株式会社	Tōkai Gyogyō K. K.	East Ocean Fisheries Company	Tōkai Fisheries	1907–1908

Table A.1: continued

Japanese Name	Romanisation	Translation	Short Reference	Dates
大日本捕鯨株式会社	Dai Nippon Hogei K. K.	Great Japan Whaling Company	Dai Nippon Whaling	1907–1909
帝国水産株式会社	Teikoku Suisan K. K.	Imperial Fisheries Company	Teikoku Fisheries	1907–1909
紀伊水産株式会社	Kii Suisan K. K.	Kii Fisheries Company	Kii Fisheries	1907–1916
長門捕鯨株式会社	Nagato Hogei K. K.	Nagato Whaling Company	Nagato Whaling	1907–1916
内外水産株式会社	Naigai Suisan K. K.	Home and Abroad Fisheries Company	Naigai Fisheries	1907–1916
大東漁業株式会社	Daitō Gyogyō K. K.	Great Easterrn Fisheries Company	Daitō Fisheries	1907–1934
土佐捕鯨合名会社	Tosa Hogei Gōmeigaisha	Tosa Whaling Unlimited Partnership	Tosa Whaling Unlimited Partnership	1907–1937
丸三製材株式会社捕鯨部	Marusan Seizai K. K. Hogeibu	Marusan Lumber Company Whaling Division	Marusan Lumber Co. Whaling Division	1908–1910
東京岩合商会捕鯨部	Tōkyō Iwaya Shōkai Hogeibu	Tokyo Iwaya Company Whaling Division	Tokyo Iwaya Co. Whaling Division	1909
大日本水産株式会社	Dai Nippon Suisan K. K.	Great Japan Fisheries Company	Dai Nippon Fisheries	1909–1916
小川島捕鯨株式会社	Ogawajima Hogei K. K.	Ogawajima Whaling Company	Ogawajima Whaling	1909–?
藤村捕鯨株式会社	Fujimura Hogei K. K.	Fujimura Whaling Company	Fujimura Whaling	1909–1934
東洋捕鯨株式会社	Tōyō Hogei K. K.	Oriental Whaling Company	Tōyō Whaling	1909–1934
遠洋捕鯨合資会社	En'yō Hogei Gōshigaisha	Pelagic Whaling Limited Partnership	En'yō Whaling Limited Partnership	1923–1943
鮎川捕鯨株式会社	Ayukawa Hogei K. K.	Ayukawa Whaling Company	Ayukawa Whaling	1925–1951
日本捕鯨株式会社	Nihon Hogei K. K.	Japan Whaling Company	Nihon Whaling	1934–1936

Japanese Name	Romanisation	Translation	Short Reference	Dates
共同漁業株式会社	Kyōdō Gyogyō K. K.	Cooperative Fisheries Company	Kyōdō Fisheries	1936–1937
北洋捕鯨株式会社	Hokuyō Hogei K. K.	North Pacific Whaling Company	Hokuyō Whaling	1936–1943
大洋捕鯨株式会社	Taiyō Hogei K. K.	Ocean Whaling Company	Taiyō Whaling	1936–1943
極洋捕鯨株式会社	Kyokuyō Hogei K. K.	Polar Oceans Whaling Company	Kyokuyō Whaling	1937–1971
林兼商店捕鯨部	Hayashikane Shōten Hogeibu	Hayashikane Firm Whaling Division	Hayashikane Shōten Whaling Division	1937–1943
日本水産株式会社	Nihon Suisan K. K.	Japan Fisheries Company	Nihon Fisheries	1937–1943
日本海洋漁業統制株式会社	Nihon Kaiyō Gyogyō. Tōsei K. K	Japan Ocean Fisheries Control Company	Nihon Kaiyō Fisheries Control Co.	1943–1945
西大洋漁業統制株式会社	Nishi Taiyō Gyogyō Tōsei K. K	Western Ocean Fisheries Control Company	Nishi Taiyō Fisheries Control Co.	1943–1945
大洋漁業株式会社	Taiyō Gyogyō K. K.	Ocean Fisheries Company	Taiyō Fisheries	1945–
日本水産株式会社	Nippon Suisan K. K.	Japan Fisheries Company	Nippon Fisheries	1945–
日東捕鯨株式会社	Nittō Hogei K. K.	Nittō Whaling Company	Nittō Whaling	1949–
日本近海捕鯨株式会社	Nihon Kinkai Hogei K. K.	Japan Coastal Whaling Company	Nihon Kinkai Whaling	1950–1970
日本小型捕鯨有限会社	Nihon Kogata Hogei Yūgengaisha	Japan Small-type Whaling Company Limited	Nihon Kogata Whaling Co. Ltd	1959–1960
北洋捕鯨有限会社	Hokuyō Hogei Yūgengaisha	Northern Oceans Whaling Company Limited	Hokuyō Whaling Co. Ltd	1960–
三洋捕鯨有限会社	San'yō Hogei Yūgengaisha	Three Oceans Whaling Company Limited	San'yō Whaling Co. Ltd	1968–
日本捕鯨株式会社	Nihon Hogei K. K.	Japan Whaling Company	Nihon Whaling	1970–
株式会社極洋	Kabushikigaisha Kyokuyō	Corporation Polar Oceans	Kyokuyō Company	1971–

Bibliography

A-Team (1992), *Kujira no kyōkun* (The moral of the whale), JMA Management Center.

Abe, Matsunoshin (1908a), 'Kanada Taiheiyōgan ni okeru hogeigyō' (The whaling industry on the Pacific coast of Canada), *Dai Nihon suisankaihō* (Journal of the Fisheries Society of Japan), 305, pp. 9–15.

Abe, Matsunoshin (1908b), 'Kanada Taiheiyōgan ni okeru hogeigyō' (The whaling industry on the Pacific coast of Canada), *Dai Nihon suisankaihō* (Journal of the Fisheries Society of Japan), 306, pp. 8–13.

Achikku myūzeamu (Attic Museum) ed. (1939), *Tosa Muroto Ukitsu gumi hogei shiryō* (Historical whaling records of the Tosa Muroto-Ukitsu group) reprinted in Nihon Jōmin Bunka Kenkyūjo (Institute for the Study of Japanese Folk Culture) ed. [1973] *Nihon jōmin seikatsu shiryō sōsho dai nijūni kan* (Collection of materials on the lives of the Japanese common people, vol. 22), San'ichi Shobō.

Akagawa, Manabu (2001), 'Gensetsu bunseki to kōchikushugi' (Discourse analysis and constructionism), in Ueno Chizuko ed. *Kōchikushugi to wa nani ka* (What is constructionism?), Keisō Shobō, pp. 63–83.

Akimichi, Tomoya (1994), *Kujira to hito no minzokushi* (An ethnography of whales and people), Tokyo Daigaku Shuppankai.

Andō, Shunkichi (1912a), 'Waga kuni ni okeru geitai no riyō' (Utilisation of the whale carcass in our country), *Dai Nihon suisankaihō* (Journal of the Fisheries Society of Japan), 355, pp. 15–21.

Andō, Shunkichi (1912b), 'Waga kuni ni okeru geitai no riyō' (Utilisation of the whale carcass in our country), *Dai Nihon suisankaihō* (Journal of the Fisheries Society of Japan), 357, pp. 27–31.

Andō, Shunkichi (1913), 'Waga kuni ni okeru geitai no riyō' (Utilisation of the whale carcass in our country), *Dai Nihon suisankaihō* (Journal of the Fisheries Society of Japan), 368, pp. 14–18.

Aomoriken Minseirōdōbu Rōseika (Division of Labour Administration, Department of Welfare and Labour, Aomori Prefecture) ed. (1969), *Aomoriken rōdō undōshi (dai ikkan).*

Asahina, Sadayoshi (1915), *Dai Nihon yōshu kanzume senkakushi: kanzume hen* (The history of western alcoholic beverages and canned foods in greater Japan: canned foods volume), Nihon wayōshu kanzume shinbunsha. Reprinted in 1997, *Meiji kōki sangyōhattatsushi shiryō, dai 339 kan* (Historical sources of late Meiji industrial development, vol. 399, Ryūkei Shosha).

Ayabe, Kazuo (1910), 'Noeruēshiki hogei ni taisuru gojin no kibō' (What we hope to get from Norwegian-style whaling), *Dai Nihon suisan kaihō* (Journal of the Fisheries Society of Japan), 335, pp. 3–4.

Baba, Komao (1942), *Hogei* (Whaling), Tennensha.

Barsh, Russel Lawrence (2001), 'Food Security, Food Hegemony, and Charismatic Animals,' in Robert L. Friedheim ed. *Toward a Sustainable*

Whaling Regime, Seattle and London: University of Washington Press, pp. 147–79.

Carwardine, Mark (1995), *Whales, Dolphins, and Porpoises*, London: Dorling Kindersley. Japanese translation, Maehata Masayoshi *et al*. trans. [1996], *Kujira to iruka no zukan*, Nihon Bōgusha.

Chōsen Gyogyō Kyōkai (Korean Fisheries Association) (1900), 'Kankai hogeigyō no ippan' (An aspect of whaling in the Korea Sea), *Dai Nihon suisankaihō* (Journal of the Fisheries Society of Japan), 212, pp. 4–19.

Chōsen Sōtokufu (Government-General of Chōsen) (1934), *Chōsen hōmotsu koseki meishō tennen kinenbutsu hozon yōmoku* (Plan for the Conservation of Treasures, Ancient Sites, Scenic Places and Natural Monuments in Chōsen).

Chōsenkai Tsūgyo Kumiai Rengōkai (Federation of Fishing Unions working in the Korea Sea) (1902a), 'Chōsenkai hogeigyō' (Whaling in the Korea Sea), *Dai Nihon suisankaihō* (Journal of the Fisheries Society of Japan), 234, pp. 24–37.

Chōsenkai Tsūgyo Kumiai Rengōkai (Federation of Fishing Unions working in the Korea Sea) (1902b), 'Chōsenkai hogeigyō' (Whaling in the Korea Sea), *Dai Nihon suisankaihō* (Journal of the Fisheries Society of Japan), 235, pp. 21–37.

Clifford, James (1988), *The Predicament of Culture*, Cambridge: Harvard University Press. Japanese transaltion, Ōta Yoshinobu *et al*. [2003], *Bunka no kyūjō*, Jinbun Shoin.

Clifford, James and George E. Marcus eds (1988), *Writing Culture*, Berkley, Los Angeles, London: University of California Press. Japanese translation, Kasuga Naoki *et al*. trans. [1996], *Bunka o kaku*, Kinokuniya Shoten.

de Certeau, Michel (1980), *L'Invention du Quotidien, 1, Arts de Faire*, Paris: Union Générale d'Editions. Japanese translation, Yamada Toyoko trans. [1987], *Nichijōteki jissen no poietiiku*, Kokubunsha. English translation, Steven Rendall trans. [1984] *The Practice of Everyday Life*, Berkeley: University of California Press.

Emi, Suiin (1907), *Jitchi tanken: hogeisen* (The whaling ship: practical exploration), Hakubunkan.

Foucault, Michel (1969), *L'Archéologie du Savoir*, Paris: Gallimard. Japanese translation, Nakamura Yūjirō trans. [1995], *Chi no kōkogaku*, Kawade Shobō Shinsha.

Freeman, Milton M. R. *et al*. (1988), *Small-type Coastal Whaling in Japan, ,* Boreal Institute for Northern Studies, the University of Alberta. Japanese translation, Takahashi Jun'ichi *et al*. trans. [1989], *Kujira no bunka jinruigaku*, Kaimeisha.

Friedheim, Robert L. (2001), 'Fixing the Whaling Regime', in Robert L. Friedheim ed. *Towards a Sustainable Whaling Regime*, Seattle and London: University of Washington Press, pp. 311–35.

Friedlander, Saul ed. (1992), *Probing the Limits of Representation*, Cambridge: Harvard University Press. Abridged Japanese translation, Uemura Tadao *et al*. trans. [1994], *Aushuvittsu to hyōshō no genkai*, Miraisha.

Fujishima, Norihito and Yoshiaki Matsuda (1998), 'IWC ni yoru geirui shigen kanri no tayōsei e no taiō ni kansuru ichi kōsatsu' (An observation concerning the response to diversity in the IWC's whale resource management), *Chiiki gyogyō kenkyū* (Regional fisheries research), 39 (1), pp. 111–24.

Fujitani, T. (1994), 'Kindai Nihon ni okeru kenryoku no tekunorojii – guntai, "chihō," shintai' (The technology of power in modern Japan – the military, 'the regions,' the body), *Shisō* (Thought), 845, pp. 163–76. (Japanese translation by Umemori Naoyuki).

Fukumoto, Kazuo (1978), *Nihon hogei shiwa* (Historical tales of Japanese whaling), Hōsei Daigaku Shuppankyoku. Reprinted [1993].

Geisharen (Federation of whale watchers) ed. (1996), *Kujira iruka zatsugaku nōto* (Miscellaneous notes on whales and dolphins), Daiyamondosha.

Hachinohe Shakai Keizaishi Kenkyūkai (Society for the study of the socio-economic history of Hachinohe) ed. (1962), *Gaisetsu Hachinohe no rekishi gekan 1* (Introduction to the history of Hachinohe vol. 2, 1), Hoppō Shunjūsha.

Hara, Takeshi (1993), *Za kujira – daigoban* (The whale – 5th edition), Bun-shindō.

Harada, Keiichi (2001), *Kokumingun no shinwa* (The myth of a popular militia), Yoshikawa Kōbunkan.

Henshūbu, Jinbunsha (Jinbunsha Editorial Department) ed. (1997), *Nihon bunken chizu chimei sōran* (A compendium of Japanese maps and place names by prefecture), Jinbunsha.

Hidemura, Senzō (1952a), 'Tokugawaki Kyūshū ni okeru hogeigyō no rōdō kankei (ichi)' (Labour relations in the whaling industry in Kyūshū during the Tokugawa period, 1), *Kyūshū Daigaku Keizai Gakkai, Keizaigaku kenkyū* (Economic Society of Kyūshū University, Journal of Political Economy), 18 (1), pp. 57–85.

Hidemura, Senzō (1952b), 'Tokugawaki Kyūshū ni okeru hogeigyō no rōdō kankei (ni)' (Labour relations in the whaling industry in Kyūshū during the Tokugawa period, 2), *Kyūshū Daigaku Keizai Gakkai, Keizaigaku kenkyū* (Kyūshū University Economics Association, Economics Research), 18 (2), pp. 67–106.

Higuchi, Hiroyoshi ed. (1996), *Hozen seibutsugaku* (Conservation biology), Tōkyō Daigaku Shuppankai.

Hokkaidō Kankyōseikatsubu Kankyōshitsu Shizenkankyōka (Natural Environment Division, Office of Environment Affairs, Department of Health and Environment, Hokkaidō Government) ed. (1998), *Dōtō chiiki ezoshika hogo kanri keikaku* (Conservation and management plan for *shika* deer in eastern Hokkaidō), Hokkaidō.

Hōshūtei, Riyū (1840), *Ogawajima keigei gassen* (The battle with the whales at Ogawajima). Reprinted in [1995] , *Nihon nōsho zenshū 58 gyogyō* (Complete collection of Japanese books on agriculture, vol 58 fishing), Nōsangyoson Bunka Kyōkai (Society for farming, mountain and fishing village culture), pp. 281–383.

Ino, Hiroya (1940), 'Suisan Nihon no sokojikara' (The underlying strength of the Japanese fishing industry), *Suisankai* (Fisheries World), 693, pp. 3–11.

Ishida, Yoshikazu (1978), *Nihon Gyominshi* (A history of Japanese fisherfolk), San'ichi Shobō.

Ishihara, Shintarō and Makoto Sataka (2000), 'Ore wa seijika to shite no sakuhin wa kaite inai' (I don't write as a politician), *Shūkan kin'yōbi* (Friday weekly), 322, pp. 17–21.

Iwasaki, Kenji (1939), 'Nanpyōyō hogei jūgunki' (Record of a whaling campaign in the Antarctic Ocean), *Suisankai* (Fisheries World), 682, pp. 53–9.

Iwatake, Mikako (1996), *Minzokugaku no seijisei*, Miraisha. (The politics of folklore).

Izukawa, Asakichi (1942), 'Kinki chūbu chihō ni okeru geiniku riyō chōsa no hōkoku gaiyō' (An overview of the report on the uses of whale meat in the Kinki and Chūbu districts), *Shibusawa Suisanshi Kenkyūshitsu Hōkoku* (Reports of the Shibusawa Fisheries Research Institute), 2, pp. 113–45. Reprinted in Nihon Jōmin Bunka Kenkyūjo (Institute for the Study of Japanese Folk Culture) ed. (1973a), *Nihon jōmin seikatsu shiryō sōsho dai ni kan* (Collection of materials on the lives of the Japanese common people, vol. 2, Tokyo: San'ichi Shobō, pp. 407–41.

Izukawa, Asakichi (1943), *Tosa hogeishi* (A history of whaling in Tosa). Reprinted in Nihon Jōmin Bunka Kenkyūjo (Institute for the Study of Japanese Folk Culture) ed. (1973c), *Nihon jōmin seikatsu shiryō sōsho dai nijūsan kan* (Collection of materials on the lives of the Japanese common people, vol. 23), San'ichi Shobō, pp. 5–703.

Kaburagi, Tokio (1932a), 'Abi toraigun'yū kaimen to gyogyō' (Flocks of migratory Pacific diver (*Gavia pacifica*) on the sea surface and fishing), in Shiseki Meishō Tennen Kinenbutsu Hozon Kyōkai (Society for the Conservation of Historical Sites Scenic Places and Natural Monuments) ed. *Tennen kinenbutsu chōsa hōkoku dōbutsu no bu dai ni shū* (Report on natural monuments – animals vol. 2), Tōkō Shoin, pp. 65–71.

Kaburagi, Tokio (1932b), 'Sunameri kujira kaiyū kaimen to gyogyō' (Migrating finless porpoises (*Neophocaena phocaenoides*) on the sea surface and fishing), in Shiseki Meishō Tennen Kinenbutsu Hozon Kyōkai (Society for the Conservation of Historical Sites Scenic Places and Natural Monuments) ed. *Tennen kinenbutsu chōsa hōkoku dōbutsu no bu dai ni shū* (Report on natural monuments – animals vol. 2), Tōkō Shoin, pp. 72–75.

Kaburagi, Tokio (1932c), 'Kabushima umineko hanshokuchi' (The Kabushima Island breeding grounds of the black-tailed gull (*Larus crassirostris*)), in Scenic Places and Natural Monuments) Shiseki Meishō Tennen Kinenbutsu Hozon Kyōkai (Society for the Conservation of Historical Sites ed. *Tennen kinenbutsu chōsa hōkoku dōbutsu no bu dai ni shū* (Report on natural monuments – animals vol. 2), Tōkō Shoin, pp. 104–7.

Kaiyō Gyogyō Kyōkai (Ocean Fishing Association) ed. (1939), *Honpō kaiyō gyogyō no gensei* (The present state of ocean fishing in Japan), Suisansha.

Kaji, Kōichi (1999), 'Hokkaidō ni okeru shika no kotaigun no kanri' (A management policy for *shika* deer populations in Hokkaidō), *Kankyō kenkyū* (Environment research), 114, pp. 78–85.

Kajimura Hideki and Kang Duk Sang (1970), 'Nitteika Chōsen no hōritsu seido ni tsuite' (On the system of law in Korea under Japanese imperialism), in Niida Noboru Hakase Tsuitō Ronbunshū Henshū Iinkai (Editorial Committee of Papers in Memory of Dr Noboru Niida) ed. *Japanese law and Asia: Papers in memory of Dr Noboru Niida, vol.3*, Keisō Shobō, pp. 319–337.

Kanda, Mikio (1981), 'Hiroshimaken no gyogyō/shoshoku' (Fishing and various trades in Hiroshima prefecture), in Kawakami Michihiko *et al* eds, *Chūgoku no seigyō: 2 gyogyō/shoshoku* (Subsistence in the Chūgoku region: 2 fishing and various trades), Meigen Shobō, pp. 181–242.

Kaneki, Jūichirō (1883), 'Hogei no chi ikani' (What of the whaling grounds?), *Dai Nihon suisankai hōkoku* (Journal of the Fisheries Society of Japan), 19, pp. 9–13.

Kasuya, Toshio (1994), 'Sunameri' (Finless porpoises (*Neophocaena phocaenoides*)), in Suisanchō (Fisheries Agency) ed., *Nihon no kishō na yasei suisei seibutsu ni kansuru kiso shiryō I* (Basic materials concerning Japan's rare aquatic wildlife I), pp. 626–634.

Kasuya, Toshio (1996), 'Handō iruka' (Bottlenose dolphin (*Tursiops truncatus*)), in Nihon Suisan Shigen Hogo Kyōkai (Japan Fisheries Resource Conservation Association) ed., *Nihon no kishō na yasei suisei seibutsu ni kansuru kiso shiryō III* (Basic materials concerning Japan's rare aquatic wildlife III), pp. 334–339.

Kasuya, Toshio and Tomio Miyashita (1994), 'Suji iruka' (Striped dolphin (*Stenella coeruleoalba*)), in Suisanchō (Fisheries Agency) ed. *Nihon no kishō na yasei suisei seibutsu ni kansuru kiso shiryō I* (Basic materials concerning Japan's rare aquatic wildlife I), pp. 616–625.

Katō, Hidehiro (1991), 'Hogei shōshi' (A brief history of whaling), in Sakuramoto Kazumi *et al* eds, *Geirui shigen no kenkyū to kanri* (Studies on whale stock management), Kōseisha Kōseikaku, pp. 264–8.

Katō, Mutsuo (1984), *Nihon no tennen kinenbutsu 1: dōbutsu I* (Japan's natural monuments 1: animals I), Kōdansha.

Kawabata, Hiroto (1995), *Kujira o totte, kangaeta* (My thoughts after killing whales), PARCO Shuppan.

Kawai, Kakuya (1924), *Zōho kaitei gyorōron* (Enlarged and revised editon: On fishing), Suisansha.

Kawashima, Shūichi and Katō Hidehiro (1991), 'Nankyokkai bosenshiki hogei hokaku tōsū to kisei no hensen' (Changes in regulations and the number of whales taken in factory-ship whaling in the Antarctic Ocean), in Sakuramoto Kazumi *et al* eds, *Geirui shigen no kenkyū to kanri* (Studies on whale stock management), Kōseisha Kōseikaku, pp. 239–55.

Kitamura, Kokujitsu (1838), *Noto no kuni saigyozue* (Illustrated compendium of fishing in Noto province). Reprinted in [1995] , *Nihon nōsho zenshū 58 gyogyō I* (Complete collection of Japanese books on agriculture, vol 58 fishing I), Nōsangyoson Bunka Kyōkai (Society for farming, mountain and fishing village culture), pp. 117–223.

Kitō, Shūichi (1996), *Shizen hogo o toinaosu* (Requestioning nature conservation), Chikuma Shobō.

Kizaki, Moritaka (1970) [1773], 'Hizenshū sanbutsu zukō – kodomo no moteasobi kujira ikken no maki' (Illustrated account of the products of Hizen province – the case of the infant playing with the whale), in Miyamoto Tsuneichi *et al* eds, *Nihon shomin seikatsu shiryō shūsei – daijikkan nōsangyomin seikatsu* (Collected historical materials on the lives of the Japanese common people vol. 10 the lives of farmers, foresters and fisherfolk), San'ichi Shobō, pp. 772–83, 818–28.

Kondō, Isao (2001), *Nihon engan hogei no kōbō* (The rise and fall of coastal whaling in Japan), San'yōsha.

Kumano Taijiura hogeishi hensan iinkai (Editorial committee of the history of whaling in Taiji bay in Kumano) ed. (1969), *Kumano Taijiura hogeishi* (A history of whaling in Taiji bay in Kumano), Heibonsha.

Kuzu, Seiichi (1932), 'Hokkaidō Matsumaegun Ōshima ni okeru ōmizunagidori hanshokuchi' (The breeding grounds of the streaked shearwater (*Calonectris leucomelas*) on Ōshima Island, Matsumaegun, Hokkaidō), in Shiseki Meishō Tennen Kinenbutsu Hozon Kyōkai (Society for the Conservation of Historical Sites, Scenic Places and Natural Monuments) ed. *Tennen kinenbutsu chōsa hōkoku dōbutsu no bu dai ni shū* (Report on natural monuments – animals vol. 2), Tōkō Shoin, pp. 8–18.

Kyokuyō Hogei sanjūnenshi henshū iinkai (Editorial committee of the thirty year history of the Kyokuyō Whaling Company)(1968), *Kyokuyō Hogei sanjūnenshi* (The thirty year history of Kyokuyō Whaling).

Maeda, Keijirō and Teraoka Yoshirō (1952), *Hogei* (Whaling), Nihon Hogei Kyōkai (Japan Whaling Association).

Marcus, George E. and Michael M. J. Fischer (1986), *Anthropology as Cultural Critique*, Chicago: University of Chicago Press. Japanese translation, Nagafuchi Yasuyuki trans. [1989], *Bunka hihan to shite no jinruigaku*, Kinonkuniya Shoten.

Marukawa, Hisatoshi (1941), *Hogeisen nikki* (Whaling ship diary), Hakubunkan.

Matsu, Makisaburō (1901a), 'Noeruēshiki hogei jikkendan' (Relating my experience of Norwegian-style whaling), *Dai Nihon suisankaihō* (Journal of the Fisheries Society of Japan),(226), pp. 11–24.

Matsu, Makisaburō (1901b), 'Noeruēshiki hogei jikkendan' (Relating my experience of Norwegian-style whaling), *Dai Nihon suisankaihō* (Journal of the Fisheries Society of Japan),(227), pp. 10–15.

Matsu, Makisaburō (1901c), 'Noeruēshiki hogei jikkendan' (Relating my experience of Norwegian-style whaling), *Dai Nihon suisankaihō* (Journal of the Fisheries Society of Japan),(228), pp. 17–22.

Matsu, Makisaburō (1901d), 'Noeruēshiki hogei jikkendan' (Relating my experience of Norwegian-style whaling), *Dai Nihon suisankaihō* (Journal of the Fisheries Society of Japan),(229), pp. 13–17.

Matsu, Makisaburō (1901e), 'Noeruēshiki hogei jikkendan' (Relating my experience of Norwegian-style whaling), *Dai Nihon suisankaihō* (Journal of the Fisheries Society of Japan),(230), pp. 18–22.

Matsuda, Motoji (1996), 'Toshi to bunka hen'yō – Shūen toshi no kanōsei' (Cities and acculturation – the possibility of rim cities), in Inoue Shun *et al* eds, *Iwanami kōza: Gendai shakaigaku 18kan Toshi to toshika no shakaigaku* (Iwanami Course: Modern sociology vol. 18 The sociology of cities and urbanisation), Iwanami Shoten, pp. 171–88.

Matsuda, Motoji (1997), 'Toshi no anākii to teikō no bunka' (Urban anarchy and the culture of resistance), in Aoki Tamotsu *et al* eds, *Iwanami kōza: Bunka jinruigaku: dairokkan, funsō to undō* (Iwanami Course: Cultural anthropology: vol. 6, distrubances and movements), Iwanami Shoten, pp. 95–134.

Matsuo, Motoyuki (1964), *Chikusan keizairon* (The economics of animal production), Ochanomizu Shobō.

Matsuura, Yoshio (1944), *Kujira* (Whales), Sōgensha.

Matsuzaki, Masahiro (1910), 'Noeruēshiki hogeigyō no hinan o benzu' (In defense of Norwegian-style whaling methods), *Dai Nihon suisankaihō* (Journal of the Fisheries Society of Japan), 337, pp. 4–7.

McAdam, Doug, John D. McCarthy and Mayer N Zald eds (1996), *Comparative Perspectives on Social Movements: Political Opportunities, Mobilizing Structures, and Cultural Framings*, Cambridge, New York and Melbourne: Cambridge University Press.

Melucci, Alberto (1989), *Nomads of the Present*, Philadelphia: Temple University Press. Japanese translation, Yamanouchi Yasushi trans. [1997], *Genzai ni ikiru yūbokumin*, Iwanami Shoten.

Melucci, Alberto (1996), *Challenging Codes*, Cambridge, New York and Melbourne: Cambridge University Press.

Mishima, Tatsuo (1899), *Hogei shinron* (A new approach to whaling), Sūzanbō.

Miyamoto, Keitarō (1940), 'Waga kuni ni genkō no kasa ni tsuite (yohō ichi)' (On headgear currently worn in Japan (preliminary report one)), *Minzokugaku Nenpō* (Annual Report of the Institute of Ethnology), 2, pp. 315–63.

Miyamoto, Keitarō (1972), 'Dai ikkan minguhen kaisetsu' (Commentary on volume one: folk artifacts), in Nihon Jōmin Bunka Kenkyūjo (Institute for the Study of Japanese Folk Culture) ed. *Nihon jōmin seikatsu shiryō sōsho dai ikkan* (Collection of materials on the lives of the Japanese common people, vol. 1), San'ichi Shobō, pp. 955–74.

Miyata, Takeshi (1959), 'Yamu o enu yūshō reppai' (Inevitable survival of the fittest), *Mainichi Shimbun*, 9th June.

Miyoshi, Manabu (1929), 'Rigaku hakase Watase Shōzaburō kun o omou' (Remembering Doctor of Science Watase Shōzaburō), *Shiseki meishō tennen kinenbutsu* (Historical sites, scenic places and natural monuments), 4, pp. 374–85.

Miyoshi, Seishi (1997), *Iruka no kureta yume* (My dolphin dream), Fuji Terebi Shuppan.

Morioka, Masahiro (1999), 'Shizen o hogo suru koto to ningen o hogo suru koto' (On protecting nature and on protecting human beings), in Kitō Shūichi ed. *Kōza ningen to kankyō dai juunikan kankyō no yutakasa o motomete* (A course on man and the environment vol. 12, towards a rich environment), Shōwadō, pp. 30–53.

Morita, Hideo (1963a), 'Nihon hogeigyō no saihensei wa hisshi ka' (Is a rearrangement of the Japanese whaling industry inevitable?), *Suisankai* (Fisheries world), 941, pp. 30–7.

Morita, Hideo (1963b), 'Nangei saihensei o meguru kokunai no ugoki' (Moves in Japan towards a rearrangement of Antarctic whaling), *Suisankai* (Fisheries world), 947, pp. 30–42.

Morita, Hideo (1965), 'Seijinshiki mukaeta Nanpyōyō hogei da ga' (Antarctic Ocean whaling comes to age but ...), *Suisankai* (Fisheries world), 971, pp. 18–31.

Morita, Katsuaki (1994), *Kujira to hogei no bunkashi* (A cultural history of whales and whaling), Nagoya: Nagoya daigaku shuppankai.

Nagasaki, Fukuzō (1984), 'Nihon no engan hogei' (Japan's coastal whaling), *Geiken tsūshin* (Whaling research newsletter), 355, pp. 75–87.

Nagasaki, Fukuzō (1990a), 'Saikin no hogei rongi ni tsuite II' (On the recent whaling debate II), *Geiken tsūshin* (The institute of cetacean research newsletter), 378, pp. 16–21.

Nagasaki, Fukuzō (1990b), 'Saikin no hogei rongi ni tsuite III' (On the recent whaling debate II), *Geiken tsūshin* (The institute of cetacean research newsletter), 379, pp. 5–7.

Naimushō Chihōkyoku (Bureau of Regional Affairs, Ministry of the Interior) ed. (1914), *Saimin chōsa tōkeihyō tekiyō* (Summary of statistics from a survey of the poor). Reprinted in abridged form in Tada Yoshizō ed. (1992), *Kakei chōsa shūsei 9 – Meiji kakei chōsashū* (Collection of surveys on household finances 9 – household finances in the Meiji period), Tokyo: Seishisha, pp. 611–20.

Naimushō Shakaikyoku Daiichibu ed. (First Department, Bureau of Social Affairs, Ministry of the Interior ed.) (1924), *Chōsenjin rōdōsha ni kansuru jōkyō* (On the situation of Korean workers) Pak Kyungsik ed. (1975) *Zainichi Chōsenjin kankei shiryō shūsei dai ikkan* (Collected materials relating to Koreans in Japan vol. 1), Tokyo: San'ichi Shobō, pp. 445–540.

Nakamura, Akira (1958), 'Shokuminchihō [hōtaisei kakuritsuki]' (Colonial law [establishment of the legal system]), in Ukai Nobushige *et al* eds, *Kōza Nihon kindaihō hattatsushi 5* (Course on the development of modern Japanese law 5), Keisō Shobō, pp. 173–206.

Nakamura, Yōichirō (1988), 'Irukaryō o megutte' (On dolphin fishing), in Shizuokaken Minzoku Geinō Kenkyūkai (Society for the Study of Folklore and Performing Arts in Shizuoka Prefecture) ed. *Shizuokaken umi no minzokushi* (Folklore of the sea in Shizuoka prefecture), Shizuoka: Shizuoka Shinbunsha, pp. 91–136.

Nakatani, Masao (1932), 'Kujira ni kansuru tenrankai ni tsuite' (On an exhibition about whaling), *Suisankai* (Fisheries world), 601, pp. 30–5.

Nakazawa, Hideo et al. (1998), 'Kankyō undō ni okeru kōgi saikuru keisei no ronri' (The dynamics of protest waves in environmental movements), *Kankyō shakaigaku kenkyū* (Environmental–sociology research), 4, pp. 142–57.

Namihira, Emiko (1978), 'Suishitai o Ebisushin to shite matsuru shinkō: sono imi to kaishaku' (The worship of a drowned body as the god Ebisu: significance and interpretation), *Minzokugaku kenkyū* (The Japanese Journal of Ethnology), 42 (4), pp. 334–55.

Nasu, Keiji (1989), 'Akishima kujira matsuri' (The Akishima whale festival), *Geiken tsūshin* (The institute of cetacean research newsletter), 375, pp. 3–8.

Nihon Honyūrui Gakkai (Mammalian Society of Japan) ed. (1997), *Reddo dēta Nihon honyūrui* (Red data Japanese mammals), Bun'ichi Sōgo Shuppan K. K.

Nihon Kajo Shuppan K.K. Shuppanbu ed. (Publishing Department, Nihon Kajo Publishing Company ed.) (1979), *Zenkoku shichōsonmei hensen sōran* (A compendium of city, town and village name changes), Nihon Kajo Shuppan K. K.

Nihon Keizai Shinbunsha (1983), *Kaisha sōkan (mijōjō kaisha ban)* (Directory of companies – unlisted company edition), Nihon Keizai Shinbunsha.

Nōrin Daijin Kanbō Tōkeika (Statistics Division, The Secretariat of the Minister of Agriculture and Forestry)(1926–30, 1932–36, 1938–42),

Daiichi – daijūhachiji nōrinshō tōkeihyō (1st to 18th tables of statistics of the Ministry of Agriculture and Forestry).

Nōrin Suisanshō Tōkei Jōhōbu: Nōrin Tōkei Kenkyūkai (Statistics and Information Division of the Ministry of Agricultue, Forests and Fisheries: Agricultural Statistics Research Unit)(1979), Suisangyō ruinen tōkei dainikan (Yearly statistics for the fishing industry vol. 2).

Nōrinshō Nōgyō Kairyōkyoku Tōkei Chōsabu (Statistics and Survey Department, Bureau of Agricultural Improvement, Ministry of Agriculture and Forestry)(1949–51), Dainijūyon – nijūrokuji Nōrinshō tōkeihyō (24th to 26th tables of statistics of the Ministry of Agriculture and Forestry), Nōrin Tōkei Kyōkai.

Nōrinshō Suisankyoku (Fisheries Bureau, Ministry of Agriculture and Forestry) ed. (1939), Hogeigyō (The whaling indistry), Nōgyō to Suisansha.

Nōsei Chōsa Iinkai (Agricultural Policy Survey Committee)(1977), Kaitei Nihon nōgyō kiso tōkei (Revised basic statistics for Japanese agriculture), Nōrin Tōkei Kyōkai.

Nōshōmu Daijin Kanbō Tōkeika (Statistics Division, Secretariat of the Minister of Agriculture and Commerce) (1925), 'Dai yonjūji nōshōmu tōkeihyō' (40th table of statistics for agriculture and commerce).

Obara, Hideo (1996), Ningen wa yasei dōbutsu o mamoreru ka (Can human beings protect wild animals?), Iwanami Shoten.

Okada, Fujie (1916), 'Kujira gyojō to shite no Nemuro kinkai no kachi' (The value of Nemuro coastal waters as a whaling ground), Suisankai (Fisheries world), 400, pp. 38–42.

Ōmura, Hideo (1938), 'Sekai hogeigyō no genjō to waga kuni hogeigyō no shōrai' (The present state of the world whaling industry and the future of Japanese whaling), Suisankai (Fisheries world), 671, pp. 14–18.

Ōmura, Hideo (1963), 'Sannin iinkai no ninmu to kokusai hogei iinkai' (The role of the special committee of 'three wise men' and the International Whaling Commission), Suisankai (Fisheries world), 941, pp. 24–8.

Ōmura, Hideo, Matsuura Yoshio and Miyazaki Ichirō (1942), Kujira – sono kagaku to hogei no jissai (Whales: Scientific facts and whaling in practice), Suisansha.

Ōno, Shishiku (1907), 'Chōshi monogatari' (Tales of Chōshi), Bungei kurabu (Literary Arts Club), 13 (9), pp. 545–60.

Ōsaka, Keikichi (1936), 'Ugokanu geigun' (The pod stayed put), Shin seinen (New youth), November. Reprinted in (2001), Ginza yūrei (Ghost on the Ginza), Tokyo Sōgensha, pp. 125–56.

Ōsumi, Seiji (1994), 'Shironagasu kujira' (The blue whale (Balaenoptera musculus)), in Suisanchō (Fisheries Agency) ed. Nihon no kishō na yasei suisan seibutsu ni kansuru kiso shiryō I (Basic materials concerning Japan's rare aquatic wildlife I), pp. 592–600.

Ōsumi, Seiji (1995), 'Kokukujira' (The grey whale (Eschrichtius robustus)), in Nihon Suisan Shigen Hogo Kyōkai ed.(Japan Fisheries Resource Conservation Association ed.), Nihon no kishō na yasei suisan seibutsu ni kansuru kiso shiryō II (Basic materials concerning Japan's rare aquatic wildlife II), pp. 513–20.

Ōta, Kōji (1927), 'Taishō jūgonendo ni okeru hogei jōkyō gaisetsu (jō)' (An overview of the whaling situation in 1926 (Part A)), Suisankai (Fisheries world), 540, pp. 20–4.

Ōtaishi, Noriyuki and Honma Hiroaki eds (1998), *Ezoshika o shokutaku e* (Bringing Hokkaidō *shika* deer to the dinner table), Maruzen Puranetto.

Oyamada, Tomokiyo (1832 [1829]), *Isanatori ekotoba* (Illustrated tales of whaling). reprinted in Miyamoto Tsuneichi *et al.* eds 1970, *Nihon shomin seikatsu shiryō shūsei daijikkan nōsangyomin seikatsu* (Collected historical materials on the lives of the Japanese common people vol. 10 the lives of farmers, foresters and fishermen), San'ichi Shobō, pp. 283–332.

Park, Koo-Byong (1995), *Jeungbopan Hanbando eonhae pogyeongsa* (Enlarged edition: A history of coastal whaling off the Korean Peninsula), Pusan: Toseochulpan Mincheok Munhwa.

Saitō, Mankichi (1918), *Nihon nōgyō no keizaiteki hensen* (Economic change in Japanese agriculture). Abridged version reprinted in Tada Yoshizō ed. (1992), *Kakei chōsa shūsei 9 Meiji kakei chōsa shū* (Collection of surveys on household finances 9 – household finances in the Meiji period), Tokyo?: Seishisha, pp. 224–54.

Sakuramoto, Kazumi (1991), 'IWC ni yoru geirui no shigen suiteichi, shigen bunrui oyobi hokaku waku' (IWC estimates of whale stocks, the classification of the resource and the whaling quota), in Sakuramoto Kazumi *et al* eds, *Geirui shigen no kenkyū to kanri* (Studies on whale stock management), Kōseisha Kōseikaku, pp. 256–61.

Sakuramoto Kazumi, Katō Hidehiro and Tanaka Shōichi eds (1991), *Geirui shigen no kenkyū to kanri* (Studies on whale stock management), Kōseisha Kōseikaku.

Satō, Ryōichi (1987), *Kujiragaisha yakiuchi jiken* (The whaling company raid incident), Saimaru Shuppankai.

Scheiber, Harry N. (1998), 'Historical Memory, Cultural Claims, and Environmental Ethics in the Jurisprudence of Whaling Regulation,' *Ocean and Coastal Management*, 38, pp. 5–40.

Sekizawa, Akekiyo (1888), 'Hogei to nishinryō no kankei ikan' (What is the connection between whaling and herring fishing?), *Dai Nihon suisankai hōkoku* (Journal of the Fisheries Society of Japan), 71, pp. 21–9.

Shigeta, Yoshiji (1962a), 'Sekai no hogei seidoshi oyobi sono haikei (san)' (The history of the world whaling system and its background (3)), *Geiken tsūshin* (Institute of Cetacean Research newsletter), 131, pp. 7–22.

Shigeta, Yoshiji (1962b), 'Sekai hogei seidoshi oyobi sono haikei (yon)' (The history of the world whaling system and its background (4)), *Geiken tsūshin* (The Institute of Cetacean Research newsletter), 133, pp. 11–20.

Shigeta, Yoshiji (1962c), 'Sekai no hogei seidoshi oyobi sono haikei (yon <jigō de go ni teisei>)' (The history of the world whaling system and its background (4 <corrected to 5 in the next issue>)), *Geiken tsūshin* (Institute of Cetacean Research newsletter), 132, pp. 5–18.

Shigeta, Yoshiji (1962d), 'Sekai no hogei seido oyobi sono haikei (roku)' (The world whaling system and its background (6)), *Geiken tsūshin* (Institute of Cetacean Research newsletter), 135, pp. 16–21.

Shigeta, Yoshiji (1962e), 'Sekai no hogei seido oyobi sono haikei (nana)' (The world whaling system and its background (7)), *Geiken tsūshin* (Institute of Cetacean Research newsletter), 136, pp. 10–17.

Shigeta, Yoshiji (1963), 'Sekai no hogei seido oyobi sono haikei (hachi)' (The world whaling system and its background (8)), *Geiken tsūshin* (Institute of Cetacean Research newsletter), 137, pp. 12–22.

Shindō, Matsuji (1985), 'Setouchi no toritsuke ajiro/sunameri ajiro' (Fishing with red-throated diver (*Gavia stellata*) and fishing with finless porpoises (*Neophocaena phocaenoides*) in the Seto Inland Sea), in Mori Kōichi ed. *Nihon minzoku bunka taikei dai jūsankan gijutsu to minzoku (jōkan)* (Compendium of Japanese folk culture vol. 13 craft and folklore [Part A]), Shōgakkan, pp. 495–5.

Shiryō Haikyū K.K. Chōsaka ed. (Stock Feed Rationing Company Research Division ed.) (1943), *Shiryō sōran* (Stock feed compendium), Shiryō Haikyū K.K.

Shizuokaken ed. (Shizuoka Prefecture ed.) (1989), *Shizuoka kenshi – Shiryō-hen 23 Minzoku ichi* (The history of Shizuoka prefecture – source materials 23 – Folklore 1), Shizuoka: Shizuokaken.

Shizuokaken ed. (Shizuoka Prefecture ed.) (1991), *Shizuoka kenshi – Shiryō-hen 23 Minzoku san* (The history of Shizuoka prefecture – source materials 23 – Folklore 3), Shizuoka: Shizuokaken.

Shizuokaken Kyōiku Iinkai Bunkaka (Cultural Affairs Division, Shizuoka Prefecture Board of Education) ed. (1986), *Shizuokaken bunkazai chōsa hōkoku dai sanjūsanshū Izu ni okeru gyorō shūzoku chōsa I* (Cultural properties of Shizuoka prefecture – report 33, Fishing folklore in Izu I), Shizuoka: Shizuokaken.

Shizuokaken Kyōiku Iinkai Bunkaka (Cultural department, Shizuoka prefecture education committee) ed. (1987), *Shizuokaken bunkazai chōsa hōkoku dai 39 shū Izu ni okeru gyorō shūzoku chōsa II* (Cultural properties of Shizuoka prefecture – report 39, Fishing folklore in Izu I), Shizuoka: Shizuokaken.

Suekawa, Hiroshi (1978), *Zentei hōgaku jiten (kaitei zōhoban)* (Fully revised dictionary of law (amended and enlarged edition)), Nihon Hyōronsha.

Tagami, Shigeru (1992), 'Kumanonada no koshiki hogei soshiki' (The old-style whaling organisation in the Kumano Sea), in Mori Kōichi ed. *Umi to rettō bunka, daihakkan: Ise to Kumano no umi* (The sea and archipelago culture, vol. 8: the seas of Ise and Kumano), Shōgakkan, pp. 369–415.

Tago, Katsuya (1926), 'Koku kujira' (The grey whale (*Eschrichtius robustus*)), *Shiseki meishō tennen kinenbutsu* (Historical sites, scenic places and natural monuments), 1 (11), pp. 1–15.

Taiyō Gyogyō hachijūnenshi hensan iinkai (Publication committee for the 80 year history of the Taiyō Fishing Company) ed. (1960), *Taiyō Gyogyō hachijūnenshi* (The eighty-year history of the Taiyō Fishing Company).

Tajima, Yoshiya (1995), 'Kaidai' (Bibliographical notes), *Nihon nōsho zenshū, 58: gyogyō ichi* (Complete collection of Japanese books on agriculture, vol. 58: fishing I), Nōsangyoson Bunka Kyōkai, pp. 384–403.

Takahashi, Jun'ichi (1987), 'Hogei no machi no chōmin aidentitii to shinboru no shiyō ni tsuite' (Local identity and use of symbols in a whaling town), *Minzokugaku kenkyū* (The Japanese Journal of Ethnology), pp. 203–212.

Takahashi, Jun'ichi (1991), 'Geirui no shigen kanri to bunka jinruigakuteki shiten no motsu igi' (The significance of an anthropological perspective on management of whale stocks), in Sakuramoto Kazumi *et al* eds, *Geirui shigen no kenkyū to kanri* (Studies on whale stock management), Kōseisha Kōseikaku, pp. 203–212.

Takahashi, Jun'ichi (1992), *Kujira no Nihon bunkashi* (A Japanese cultural chronicle on whales), Tankōsha.

Takubo, Yūko (1997), 'Makimachi "jūmin tōhyō o jikkō suru kai" no tanjō, hatten to seikō' (The emergence, development and success of the 'Referendum Advocacy Association' in Maki), *Kankyō shakaigaku kenkyū* (Journal of Environmental Sociology), 3, pp. 131–48.

Tennensha Jiten Henshūbu (The Dictionary Editorial Department of Tennensha) ed. (1963), *Senpaku jiten* (Dictionary of shipping), Tennensha.

Thompson, E. P. (1980), *The Making of the English Working Class*, Harmondsworth: Penguin Books. Japanese translation, Ichihashi Hideo and Haga Ken'ichi trans. [2003], *Ingurando rōdōsha kaikyū no keisei*, Seikyūsha.

Torigoe, Hiroyuki (1997), *Kankyō shakaigaku no riron to jissen* (Environmental sociology in theory and practice), Yūhikaku.

Torisu, Kyōichi (1999), *Saikai hogei no shiteki kenkyū* (Historical research on whaling in the seas around Kyūshū), Kyūshū Daigaku Shuppankai.

Touraine, Alain *et al.* (1980), *La Prophétie Anti-nucléaire*, Paris: Editions du Seuil. Japanese translation, Itō Ruri trans. [1984], *Hangenshiryoku undō no shakaigaku*, Shinsensha.

Tōyō Hogei K.K. (Tōyō Whaling Company) ed. (1910), *Honpō no Noeruēshiki hogeishi* (A history of Norwegian-type whaling in Japan).

Tsuro Hogei K.K. (Tsuro Whaling Company)(1902), *Tsuro hogeishi* (An account of Tsuro whaling).

Uchida, Seinosuke (1925), 'Kōchiken Birōjima ōmizunagidori hanshokuchi' (The breeding grounds of the streaked shearwater (*Calonectris leucomelas*) on Birōjima island, Kōchi prefecture), in Uchida Seinosuke and Kuroda Nagamichi eds, *Tennen kinenbutsu chōsa hōkoku dōbutsu hen dai isshū* (Report on a survey of natural monuments: animals vol. 1.), Hakuhōsha, pp. 83–98.

Ueno, Chizuko (1997), 'Kioku no seijigaku' (The politics of memory), *Inpakushon* (Impaction), 103, pp. 154–74.

Victor, David G. (2001), 'Whale Sausage: Why the Whaling Regime Does Not Need to be Fixed,' in Robert L. Friedheim ed. *Toward a Sustainable Whaling Regime*, Seattle and London: University of Washington Press, pp. 292–310.

Watanabe, Hiroyuki (2000), 'Watase Shōzaburō no shizenkan – seibutsu no inyū to tennen kinenbutsu no seitei/shitei o megutte' (Watase Shōzaburō's view of nature – centering on the introduction of organisms and the establishment and the designation of natural monuments), *Kagakushi kenkyū* (Journal of the history of science, Japan), 213, pp. 1–10.

Watase, Shōzaburō (1916a), 'Kitsune no yōshoku (ichi)' (Fox farming (1)), *Tōkyō nichinichi shinbun*, 30 October.

Watase, Shōzaburō (1916b), 'Kitsune no yōshoku (ni)' (Fox farming (2)), *Tōkyō nichinich shinbun*, 31 October.

Watase, Shōzaburō (1916c), 'Kitsune no yōshoku (san)' (Fox farming (3)), *Tōkyō nichinich shinbun*, 3 November.

Watase, Shōzaburō (1916d), 'Kitsune no yōshoku (yon)' (Fox farming (4)), *Tōkyō nichinichi shinbun*, 4 November.

Watase, Shōzaburō (1916e), 'Kitsune no yōshoku (go)' (Fox farming (5)), *Tōkyō nichinichi shinbun*, 5 November.

Watase, Shōzaburō (1916f), 'Kitsune no yōshoku (roku)' (Fox farming (6)), *Tōkyō nichinich shinbun*, 6 November.

Watase, Shōzaburō (1916g), 'Kitsune no yōshoku (nana)' (Fox farming (7)), *Tōkyō nichinichi shinbun*, 7 November.

Watase, Shōzaburō (1921a), 'Shizenkai no fukkyū jigyō (jō)' (The task of restoring the natural world (part A)), *Tōkyō nichinichi shinbun*, 27 September.

Watase, Shōzaburō (1921b), 'Shizenkai no fukkyū jigyō (chū)' (The task of restoring the natural world (part B)), *Tōkyō nichinichi shinbun*, 28 September.

Watase, Shōzaburō (1921c), 'Shizenkai no fukkyū jigyō (ge no jō)' (The task of restoring the natural world (part C-1)), *Tōkyō nichinichi shinbun*, 29 September.

Watase, Shōzaburō (1921d), 'Shizenkai no fukkyū jigyō (ge no jō)' (The task of restoring the natural world (part C-2)), *Tōkyō nichinichi shinbun*, 30 September.

Watase, Sadao (1965), 'Engan hogei no gōrika mondai to kongo no kadai' (The question of rationalisation of coastal whaling and issues for the future), *Suisankai* (Fisheries world), 962, pp. 39–49.

Watase, Sadao (1995), 'Hogei hiwa' (Secret tales of whaling), *Ima da kara hanasō chinmoku no jikō* (Now let's talk about it – the reign of silence is over), Seisei Shuppan, pp. 235–240.

Yamaguchi, Kazuo (1973), 'Dai nijūsankan Chūgoku-Shikokuhen (4) Kaisetsu' (Vol. 23 Chūgoku and Shikoku (4) Commentary), in Nihon Jōmin Bunka Kenkyūjo (Institute for the Study of Japanese Folk Culture) ed. *Nihon jōmin seikatsu shiryō sōsho dai nijūsankan* (Collection of materials on the lives of the Japanese common people, vol. 23), San'ichi Shobō, pp. 835–42.

Yamashita, Shōto (2004a), *Hogei I* (Whaling I), Hōsei Daigaku Shuppan-kyoku.

Yamashita, Shōto (2004b), *Hogei II* (Whaling II), Hōsei Daigaku Shuppan-kyoku.

Yasumaru, Yoshio (1965a), 'Nihon no kindaika to minshū shisō (jō)' (Japanese modernisation and popular thought (Part A)), *Nihonshi kenkyū* (Journal of Japanese history), 78, pp. 1–19.

Yasumaru, Yoshio (1965b), 'Nihon no kindaika to minshū shisō (ge)' (Japanese modernisation and popular thought (Part B)), *Nihonshi kenkyū* (Journal of Japanese history), 79, pp. 40–58.

Yatsu, Naohide (1931), 'Watase hakase ryakuden' (A short biography of Dr Watase), *Dōbutsugaku zasshi* (Zoological magazine), 508/509, pp. 45–46.

Yoshida, Ryūji (1994), 'Gunshū kōdō to nichijōsei – Kamagasaki daiichiji bōdō o jirei to shite' (Crowd behaviour and everyday life – the case of the first Kamagasaki riot), *Soshioroji* (Sociology), 39 (2), pp. 75–95.

Yoshioka, Kōkichi (1938), 'Tosa Muroto Ukitsugumi hogei jitsuroku' (Actual whaling records of the Tosa Muroto Ukitsu guild), in Nihon Jōmin Bunka Kenkyūjo (Institute for the Study of Japanese Folk Culture) ed., *Nihon jōmin seikatsu shiryō sōsho dai nijūnikan* (Collection of materials on the lives of the Japanese common people, vol. 22), San'ichi Shobō, pp. 393–487.

Yoshioka, Shōzō, Ōnishi Nobuhiko and Takano Morio (1963), *Yōkei kōza I* (A course on poultry farming I), Asakura Shoten.

Newspapers, magazines and other sources where the author's name is not recorded

Chapter 1

'*Isanatori ekotoba*' (Illustrated tales of whaling) (1832), in Miyamoto Tsuneichi *et al.* eds 1970, *Nihon shomin seikatsu shiryō shūsei dai jikkan nōsangyomin seikatsu* (Collected historical materials on the lives of the Japanese common people vol. 10 the lives of farmers, foresters and fisherfolk), Tokyo: San'ichi Shobō, pp. 283–332.

'Nichiro ryōkokujin no Kankai hogei jōkyō' (The state of Japanese and Russian whaling in the Korean Sea) (1904), *Dai Nihon suisankaihō* (Journal of the Fisheries Society of Japan), 260, pp. 34–6.

'Sekizawa Akekiyo kun no den' (Sekizawa Akekiyo's biography) (1897a), *Dai Nihon suisankaihō* (Journal of the Fisheries Society of Japan), 178, pp. 33–7.

'Sekizawa Akekiyo kun no den' (Sekizawa Akekiyo's biography) (1897b), *Dai Nihon suisankaihō* (Journal of the Fisheries Society of Japan), 179, pp. 38–40.

'Tōyō Hogei K.K. nijūyoman'en shūeki zōka yosō' (Over two hundred thousand yen predicted increase in profits for the Tōyō Whaling Company) (1926), *Ōsakaya Shōten Junpō* (Osakaya Company ten-day reports), 135, pp. 8–10.

Chapter 2

'Dotō no hibiki' (The sound of the angry waves) (1910), *Dai Nihon suisankaihō* (Journal of the Fisheries Society of Japan), 339, pp. 32–3.

'Gyomin bōdō yobun' (What they're saying about the fishermen's riots) (1911), *Tōō Nippō* (Tōō Daily), 11 November.

'Gyomin Shaonkai' (Fishermen's Thanksgiving Meeting) (1912), *Ōnan Shinpō* (Ōnan News), 19 October.

'Hogei bōdō jiken shokan' (Impressions of the whaling riot incident) (1911), *Tōō Nippō* (Tōō Daily), 12 December.

'Hogei hantoshi no kansatsu' (Half a year's observation of whaling) (1911), *Hachinohe*, 1 October.

'Hogeigaisha yakiuchi jiken zokuhō' (A further report on the whaling company raid incident) (1911), *Hachinohe*, 7 November.

'Jūōin no Hōtōkai' (The Memorial Service at the Jūōin) (1912), *Hachinohe*, 19 October.

'*Kujiragaisha yakiuchi jiken kōhan kiroku*' (Record of the public trial over the whaling company raid incident).

'"Nanpyōyō imin" no kiroku' (An account of the 'nomads of the Antarctic Ocean') (1957), *Suisan Kikan* (Fisheries Quarterly), 4, pp. 96–9.

'Same gyomin bōdō shōhō' (Detailed report on the fishermen's riots in Same) (1911), *Tōō Nippō* (Tōō Daily), 3 November.

'Same bōdō jiken kōhan' (Public trial over the fishermen's riot incident in Same) (1912), *Tōō Nippō* (Tōō Daily), 6 February.

'Same bōdō yoshin shūketsu' (End of the preliminary hearing over the riots in Same) (1911a), *Ōnan Shinpō* (Ōnan News), 10 December.

'Same bōdō yoshin shūketsu' (End of the preliminary hearing over the riots in Same) (1911b), *Ōnan Shinpō* (Ōnan News), 13 December.

'Shaon no jitsu o ageyo (ichi)' (Offer up the fruits of gratitude (1)) (1912), *Hachinohe*, 19 October.

'Shaon no jitsu o ageyo (ni)' (Offer up the fruits of gratitude (2)) (1912), *Hachinohe*, 22 October.

'Takagi kenji no kunwa (ichi)' (The lesson of prosecutor Takagi (1)) (1912), *Ōnan Shinpō* (Ōnan News), 19 October.

'Takagi kenji no kunwa (ni)' (The lesson of prosecutor Takagi (2)) (1912), *Ōnan Shinpō* (Ōnan News), 22 October.

'Takagi kenji no kunwa (san)' (The lesson of prosecutor Takagi (3)) (1912), *Ōnan Shinpō* (Ōnan News), 25 October.

'Tōyō Hogeigaisha de Same jigyōba o yasumu' (Tōyō whaling stops work at its Same station) (1932), *Ōnan Shinpō* (Ōnan News), 13 March.

Chapter 3

Chōsen Sōtokufu kanpō (Official gazette of the Government-General of Chōsen) (1985–8), edited by Hanguk Hakmun Munhyeon Yeon'gujang (Korean Institute for Research into Academic Sources), Seoul: Asea Munhwasa.

Kanpō (Official gazette) No. 7899 (21 October 1909); No. 2258 (16 February 1920); No. 2245 (27 June 1934); No. 2269 (25 July 1934); No. 3427 (8 June 1938).

Chōsen Sōtokufu Kanpō (Official gazette of the Government-General of Chōsen) No. 1572 (1 November 1917); No. 2611 (27 April 1921); No. 3092, (1 December 1922); extra edition (10 December 1929); extra edition, (8 August 1933); No. 4612 (15 June 1942); No. 5136 (20 March 1944); No. 5315 (21 October 1944).

Chapter 4

'Geiniku o uru' (Whale meat to go on sale) (1919), *Tōkyō Nichinichi Shinbun*, 26 July.

'Shokuryōhin no hippaku to geiniku shokuyō' (Scarcity of foodstuffs and the consumption of whale meat) (1919), *Suisankai* (Fisheries world), 444, pp. 35–6.

Chapter 5

'Dai jūniji (mama) nangei no shutsuryō keikaku' (12th (sic.) Antarctic Ocean whaling plan) (1964), *Suisankai* (Fisheries world), 959, pp. 48–56.

'Dai rokujukkai shōshūkai yōroku' (Summary of the sixtieth small group meeting) (1888), *Dai Nihon suisankai hōkoku* (Journal of the Fisheries Society of Japan), 70, p. 1.

'Kankeikoku chūshi no naka o Nihon sendan wa shutsuryō shita' (The Japanese fleet puts to sea under the gaze of member countries) (1963), *Suisankai* (Fisheries world), 947, pp. 43–52.

'Meidō shita hogei gyōkai' (Whaling industry thrown into confusion) (1957), *Suisankai* (Fisheries world), 876, pp. 20–8.

Suisankai (Fisheries world), numbers 745–1068.

Kanpō (Official gazette) No. 6269 (1947).

Websites

Prologue

'Dai 55kai Kokusai Hogei Iinkai (IWC) nenji kaigō kekka' (Results of the 55th annual meeting of the IWC), *http*://www.jfa.maff.go.jp/release/15.07.07.1.html (accessed on 3 November 2004).

'Dai 57kai Kokusai Hogei Iinkai (IWC) nenji kaigō sōkai no kaisai ni tsuite' (On the opening of the general assembly of the 57th annual meeting of the IWC), *http://www.jfa.maff.go.jp/release/17/050617IWCannualstart.pdf* (accessed on 1 August 2005).

'Dai 57kai Kokusai Hogei Iinkai (IWC) nenji kaigō sōkai no kekka ni tsuite' (On the results of the general assembly of the 57th annual meeting of the IWC), *http://www.jfa.maff.go.jp/release/17/17.0624.02htm* (accessed on 1 August 2005).

'Dai 56kai Kokusai Hogei Iinkai (IWC) nenji kaigō honkaigō no kekka ni tsuite' (On the results of the main conference of the 56 th annual meeting of the IWC), *http://www.jfa.maff.go.jp/release/15.07.07.1.html* (accessed on 3 November 2004).

'Dai 54kai Kokusai Hogei Iinkai (IWC) nenji kaigō kekka' (Results of the 54th annual meeting of the IWC), *http://www.jfa.maff.go.jp/release/14.05.31.7.html* (accessed on 3 November 2004).

'Geirui no hokaku nado o meguru naigai no jōsei heisei jūgonen shichigatsu' (Internal and external conditions surrounding the taking of whales etc, July 2003), *http://www.jfa.maff.go.jp/whale/document/brief_explanation_of_whaling_jp.htm* (accessed on 3 November 2005).

'Hogeihan no kihonteki na kangaekata' (The fundamental thinking of the whaling section), *http://www.jfa.maff.go.jp/whale/assertion/assertionjp.htm* (accessed on 1 August 2005).

'Kogata hogeigyō ni kansuru kiso chishiki' (Basic facts about the small type whaling industry), *http://homepage2.nifty.com/jstwa/kisochisiki.htm* (accessed on 1 August 2005).

'Revised Management Scheme' *http://www.iwcoffice.org/conservation/rms.htm* (accessed on 1 August 2005).

'Taxonomy of Whales' *http://www.iwcoffice.org/conservation/cetacea.htm* (accessed on 1 August 2005).

'2005 Meeting' *http://www.iwcoffice.org/meetings/meeting2005.htm* (accessed on 1 August 2005).

Chapter 3

'Ezoshika no hogo to kanri' (The conservation and management of Hokkaidō *shika* deer), *http://www.pref.Hokkaidō.jp/kseikatu/ks-kskky/sika/si katop.htm* (accessed on 22 September 2004).

Chapter 5

Kokusai Hogei Iinkai (IWC) tokubetsu kaigō no kekka ni tsuite' (On the results of the special meeting of the IWC), *http://www.jfa.maff.go.jp/release/14.10.15.4.htm* (accessed on 15 October 2004).

Index